Naval gunnery; a description & history of the fighting equipment of a man-of-war

H Garbett

ROYAL NAVY HANDBOOKS

EDITED BY

COMMANDER C. N. ROBINSON, R.N.

Crown 8vo, 5s. each.

1. **Naval Administration.** By ADMIRAL SIR R. VESEY HAMILTON, G.C.B.
2. **The Mechanism of Men-of-War.** By FLEET-ENGINEER REGINALD C. OLDKNOW, R.N.
3. **Torpedoes and Torpedo-Vessels.** By LIEUTENANT G. E. ARMSTRONG, late R.N.
4. **Naval Gunnery, a Description and History of the Fighting Equipment of a Man-of-War.** By CAPTAIN H. GARBETT, R.N.

The following volumes are in preparation.

5. **The Entry and Training of Officers and Men of the Royal Navy and the Royal Marines.** By LIEUT. J. N. ALLEN, late R.N.
6. **Naval Strategy and the Protection of Commerce.** By PROFESSOR J. K. LAUGHTON, R.N.
7. **The Internal Economy of a Man-of-War.**
8. **Naval Architecture.**
9. **Dockyards and Coaling Stations.**
10. **Naval Tactics.**
11. **Naval Hygiene.**
12. **The Laws of the Sea.**

LONDON: GEORGE BELL AND SONS.

ROYAL NAVY HANDBOOKS

EDITED BY

COMMANDER CHARLES NAPIER ROBINSON, R.N.

NAVAL GUNNERY

GEORGE BELL & SONS
LONDON: YORK STREET, COVENT GARDEN
NEW YORK: 66, FIFTH AVENUE, AND
BOMBAY: 53, ESPLANADE ROAD
CAMBRIDGE: DEIGHTON, BELL & CO.

West & Co., Southsea, photo. *Frontispiece.*

NAVAL GUNNERY

A DESCRIPTION AND HISTORY OF THE
FIGHTING EQUIPMENT OF A
MAN-OF-WAR

BY
CAPTAIN H. GARBETT, R.N.

LONDON
GEORGE BELL AND SONS
1897

CHISWICK PRESS —CHARLES WHITTINGHAM AND CO
TOOKS COURT, CHANCERY LANE, LONDON

TO THE

LARGELY INCREASING NUMBERS OF THE GENERAL PUBLIC, WHO

ARE INTERESTING THEMSELVES IN NAVAL MATTERS,

THIS LITTLE WORK IS RESPECTFULLY

DEDICATED BY THE

AUTHOR.

PREFACE.

THIS work has been written, not for experts, but to give the general public some information on the all-important subject of the armament of our ships of war, which they may not be in a position to obtain conveniently for themselves. There is much that is not original in the book, as it is, from its very nature, a compilation in the main from the standard works and text-books on gunnery, which are not as a rule available for use by the man in the street. My aim has been, therefore, to trace the history of naval gunnery from the date when guns are first mentioned as having been used on board ships down to our own time, and to put into a readable form enough of the heavy matter contained in the standard text-books to give non-professional readers a fair insight into the causes which have brought us from the smooth-bore muzzle-loading 68-pounder, the heaviest gun in existence at the time of the Russian War, to the breech-loading 111-ton guns of the *Sans Pareil* and *Benbow*, and from the smooth-bore 32-pounder of the same period to the 6-inch quick-firing gun of to-day; and also to give some idea of how the guns on board men-of-war are now constructed, mounted and worked, and how complex but formidable a fighting machine is the battleship of the present day, when compared to the wooden ships of the line and frigates of only thirty years ago. Great as have been the changes and improvements in everything connected with naval gunnery during the last thirty years, there is as yet no appearance of finality; every new ship sees some improvement effected in the mounting or method of working her guns, such as the substitution of electrical for hydraulic power in some of our newest ships, so that it is impossible that any book on gunnery can be quite up to date; I have, however, added as far as possible the accounts of improvements, gathered from the reports of the gun trials of our latest ships, which have appeared in the columns of the "Times" and other sources.

I am greatly indebted to Commander E. Lloyd, R.N., and Mr. Hadcock, for their kind permission to reproduce any plates I wished from their work on "Artillery, its Progress, and Present Position," and to avail myself of information contained in that book; much of the details in regard to the hydraulic and Vavasseur mountings and quick-firing guns have been compiled from that valuable work. My thanks are due also to the Inspector-General of Ordnance for permission to reproduce from the Service Text-books, also to the Maxim and Hotchkiss Companies for the same permission, while I have received much valuable assistance from Captain H. Grenfell, R.N., Commander A. E. A. Grant, R.N., Captain Prinsep, R.A., and Lieutenant P. N. Wright, R.N. The profiles and deck plans of ships in the chapter on "Battleships" have been reproduced from the "Marine Almanach," permission to do so having been most courteously accorded me by the Austrian Imperial Hydrographical Department at Pola.

THE AUTHOR.

CONTENTS.

CHAP.		PAGE
I.	Naval gunnery up to the period of the Russian War	1
II.	The first introduction of rifled guns—the Armstrong breech-loading and other systems	31
III.	The adoption of muzzle-loading rifled guns, their construction, rifling, mounting, etc.	47
IV.	General remarks on breech-loading rifled guns, from their re-introduction in 1879	79
V.	Remarks on the construction of breech-loading guns (continued)	96
VI.	The different natures of breech-loading guns in the Navy, their characteristics, etc.	112
VII.	Naval gun-mountings of the present day	135
VIII.	Quick-firing guns	159
IX.	Quick-firing guns (continued)	177
X.	Magazines, shell-rooms, ammunition supply to guns, and loading arrangements	188
XI.	Sights, the director, firing arrangements, and conning tower	201
XII.	Powder and cordite	215
XIII.	Projectiles	229
XIV.	Fuzes	263
XV.	The development of the modern battleship . . .	294
XVI.	The fighting organization of a battleship, and the men who fight the guns	328
XVII.	Concluding remarks	347

LIST OF ILLUSTRATIONS.

PLATES.

		TO FACE PAGE
1	H.M.S. *Majestic.* Flag-ship of Channel Squadron . *Frontispiece*	
2.	H.M.S. *Victoria* firing one of her 111-ton Guns, from an instantaneous photograph	64
3.	Quarter-deck of H.M S. *Royal Sovereign*, showing 67-ton Guns in after Barbette	97
4.	H.M.S. *Benbow*, showing 111-ton Gun in after Barbette .	122
5.	Quarter-deck of H M.S *Royal Sovereign*, looking forward, showing after Barbette and Lower Fighting Top .	158
6.	6-Pounder Quick-firing Gun. No. 1 Laying the Gun .	170
7.	6-inch Quick-firing Gun on Upper Deck Mounting, secured inboard	176
8.	6-inch Quick-firing Gun on Upper Deck Mounting, Breech Open	214
9.	Company of Bluejackets at Cutlass Drill. "Point"	240
10.	Battalion Drill Forming Square	294
11.	Field-Gun Drill	320
12	Cutlass Drill on Quarter-deck of *Royal Sovereign.* "Slope Swords"	340

ILLUSTRATIONS IN THE TEXT.

FIG		PAGE
1.	Gun recovered from Wreck of the *Mary Rose;* temp. Henry VIII.	4
2.	Petrieroe a Braga and the Moscolo	8
3.	32-Pounder Carronade	15
4.	Hidden's Gun-Lock	21

FIG PAGE
5. Brass Tangent Sight 24
6. 32-Pounder Gun, Working with Reduced Crew of Nine Men. "Elevate". 25
7. Marked Quoin and Bed 26
8. 32-Pounder Gun's Crew, Close Up 27
9. 32-Pounder Gun's Crew. Extreme Train "Right". . 28
10. ,, ,, ,, ,, ,, "Left" . 29
11. 40-Pounder Armstrong Gun 35
12. Polygroove Rifling of Early Armstrong Guns . . . 36
13. Vent-Piece. 39
14. 40-Pounder Breech-Screw, Tappet Ring, and Levers . 39
15. Armstrong Barrel-Headed Sights 42
16. The Shunt Groove 51
17. The Plain Groove 52
18. 64-Pounder Gun. Earliest Pattern 52
19. ,, ,, Latest Pattern 53
20. The Woolwich Groove 53
21. Maitland Muzzle-loading Groove 55
22 Section of 80-ton Gun, showing Method of Building Up . 57
23. Breech-piece in Coil and ready for shrinking . . . 59
24. Trunnion Ring 60
25. Trunnion Ring and Jacket 60
26. The Cascable 61
27. 100-Pounder Smooth-Bore Gun, Mounted on Wooden Carriage and Slide. 67
28. Wooden Slide for Heavy Guns 67
29. Compressor 68
30. Iron Carriage and Slide for 7-inch and 9-inch Guns. . 69
31. Upper Face of Slide 70
32. Cross Section of Carriage, showing Elswick Plate Compressor 71
33. Iron Carriage for 7-inch Gun 72
34. 18-ton Gun on Scott Broadside Mounting 74
35. The Bow Compressor, Invented by Rear-Admiral Scott . 75
36. 25-ton Muzzle-loading Gun, as Mounted in Turret of H.M.S. *Glatton* 77
37. The "Cup" Obturator 86
38. The De Bange Pad and Obturator 87
39. Breech Jacket and Screw 92
40. 9·2-inch 22-ton Gun 116

FIG. PAGE
41. The 16 5-inch 111-ton Gun 122
42. 6-inch Hand-worked Breech Mechanism. Woolwich Design 127
43. 9·2-inch and 10-inch Hand-worked Breech Mechanism. Woolwich Design 129
44. Hydraulic Breech Mechanism for 67-ton Guns. End View 132
45. Hydraulic Breech Mechanism for 67-ton Guns, Section . 133
46. 6-inch Breech-loading Gun on Vavasseur C. P. Mounting 145
47. Live Roller Ring 148
48. Turret of H.M.S. *Thunderer*, showing Mounting of 10-inch Breech-loading Gun 150
49. Hydraulic Mounting for Turret Gun, showing Recoil and Elevating Presses 155
50. Maxim Gun on Naval Mounting 161
51. Maxim Gun 162
52. Breech Mechanism of Hotchkiss 3-Pounder Quick-firing Gun 166
53. 6-Pounder Quick-firing Gun on Elswick Automatic Recoil Mounting 169
54. 6-inch Elswick Quick-firing Gun, Breech open, ready for Loading 175
55. Section of 6-inch Quick-firing Wire Gun 176
56. Electrical and Percussion Firing Gear for Quick-firing Guns 181
57. 4·7-inch Quick-firing Gun 182
58. 6-inch Quick-firing Gun on Centre Pivot Pedestal Mountings 185
59. Elswick New Breech Mechanism for Quick-firing Ordnance 186
60. Cylindrical Cartridge Case 191
61. "A" Cartridge Case. . . . 191
62. Hydraulic Rammer 194
63. H Speed Sight . . . 202
64. Fore Sight 202
65. Side View of Fore Sight 202
66. "Grenfell" Night Sight 207
67 *Ibid.* 208
68. Powders and Cordite 224
69. Chain and Bar Shot 231

FIG PAGE
70. Caffin's Grape Shot 232
71 Boxer Shrapnel. First Pattern 236
72. Diaphragm Shrapnel Shell 237
73. Carcase 238
74. Ground Light Ball 238
75. Parachute Light Ball 239
76. Armstrong 12-pounder Case Shot 244
77. Armstrong Segment Shell 245
78. Illustrating Studs for Increasing Twist of Rifling . . 247
79. Common Shell, Studless, 10-inch R. M. L. 249
80. Shrapnel Shell, Studless, 10-inch R. M. L. 250
81. Palliser Shot, B. L , 12-inch Section 253
82. Case Shot, B. L., 13·5-inch Section 255
83. Shrapnel Shell, B. L., 5-inch Section 255
84. Cartridge, Nordenfelt Quick-firing 6-Pounder . . 258
85. ,, S. A. Ball, Martini-Henry 258
86. ,, S. A. Ball, Magazine Rifle, Cordite . . 259
87. 24-Pounder War Rocket 261
88. Large Mortar Time Fuze 268
89. 15-Second Wood Time Fuze 270
90. ,, ,, ,, ,, with Detonator Section . 270
91. Armstrong E. Time Fuze 271
92. Sensitive Time Fuze, "Middle" 273
93. Moorsom Percussion Fuze. Section 275
94. Pettman's G. S. Percussion Fuze 278
95. R. L. Percussion Fuze 280
96. Fuze, Direct Action 281
97. Fuze, Percussion, Small 283
98. Hotchkiss Base Percussion Fuze 284
99. Nordenfelt Percussion Base Fuze 285
100. Armstrong Base Fuze 286
101. (Medium) Armstrong Time and Concussion Fuze . . 288
102. Profile and Deck Plans, showing Disposition of Armour and Guns, of H.M.S. *Bellerophon* 301
103. Profile and Deck Plans, showing Disposition of Armour and Guns, of H.M S. *Alexandra* 303
104. Profile and Deck Plans, showing Disposition of Armour and Guns, of H M S. *Devastation* 305
105. Profile and Deck Plans, showing Disposition of Armour and Guns, of H.M.S. *Colossus* 307

FIG. PAGE

106. Profile and Deck Plans, showing Disposition of Armour and Guns, of "Admiral" class 309

107. Profile and Deck Plans, showing Disposition of Armour and Guns, of H M.S. *Nile* and *Trafalgar* . . . 311

108. Profile and Deck Plans, showing Disposition of Armour and Guns, of *Royal Sovereign* class 313

109. Profile and Deck Plans, showing Disposition of Armour and Guns, of *Majestic* class 315

110. Profile and Deck Plans, showing Disposition of Armour and Guns, of French battleships *Amiral-Baudin* and *Formidable* 319

111. Profile and Deck Plans, showing Disposition of Armour and Guns, of French battleships of *Neptune* class . 321

112. Profile and Deck Plans, showing Disposition of Armour and Guns, of French battleships of *Bouvet* class . 323

113. Profile and Deck Plans, showing Disposition of Armour and Guns, of French battleships of *Charlemagne* class 325

NAVAL GUNNERY.

CHAPTER I.

NAVAL GUNNERY UP TO THE PERIOD OF THE RUSSIAN WAR.

JOHN BARBOUR, Archdeacon of Aberdeen, who wrote in the year 1375, speaks of cannon, under the name of "Crakys of War," as having been employed by Edward III. in the Scotch campaign of 1327, while existing records furnish convincing evidence that the king carried cannon with him in his expedition to France in 1346, and that guns were mounted in at least some of the ships forming his fleet. In the Public Record Office[1] is to be found an "Indenture between John Starlyng, formerly Clerk of the Ships, Galleys, Barges, Balingers and others the King's Vessels and Helmyng Leget Keeper of the same," bearing date, 22nd June, 12 Edward III., 1358, supplying to the latter, among certain other stores for the ship called the *Bernard de la Tour,* two iron guns, also to the *Marie de la Tour,* one iron gun of two chambers and another of brass of one chamber. Cannon were certainly used in a sea-fight between the Moors of Tunis and of Seville in 1350; while the Venetians employed artillery against the Genoese at sea in 1377, but the guns of those early days were undoubtedly very small, and, until the latter end of the fourteenth century, the

[1] Roll T. G. 11,097; "History of the Royal Navy," by Sir N. H. Nicolas.

balls probably did not exceed 2 lb. or 3 lb. in weight, many of the guns being, curiously enough, breech-loading.

During the Wars of the Roses the development of the fleet fell into abeyance, and for all practical purposes the history of naval gunnery need not be considered before the latter part of the fifteenth century, as it is only from that period that we have any authentic record with regard to the nature and number of the guns forming the armament of the few large ships of the fleet of that day.

Many naval writers refer to a ship called the *Great Harry*, which is supposed to have been built by order of Henry VII., but, according to Mr. Oppenheim, there is no trace of such a ship in the State Papers. Two new ships, the *Regent* and the *Sovereign*, were, however, built about the year 1488 by Henry's orders, vessels of 600 tons, carrying an armament of 180 guns, most of which were serpentines, only throwing a 12-oz. ball. The *Sovereign* was rebuilt in 1509, when the number of her guns was reduced to seventy-one, but the size of a few of them was much increased, as she was now made to carry four whole and three half curtalls of brass, three culverins, two falcons, and eleven heavy iron guns as part of her armament.[1] The curtall or curtow, a 60-pounder, which had hitherto only been used as a siege gun on land, was probably the "cannon" of Sir W. Monson (see p. 6), as the term "cannon," not "curtall," is used in an original manuscript (see p. 5) giving the description of the *Henry Grace-à-Dieu*, also built at that time; and its being mounted on board ships marks a great advance in their armament. Port-holes also by this time had come into general use, their introduction being generally attributed to Descharges, a French shipwright, of Brest, in 1500, but they were certainly known before that date, as both the *Regent* and *Sovereign* had port-holes, although their general adaptation for the purpose of

[1] "The Administration of the Royal Navy, 1509-1660." M. Oppenheim.

broadside fire was doubtless one of the improvements made in ships of war at this period.

According to Charnock,[1] the original name of the *Regent* was *Great Harry*, the name being changed on the accession of Henry VIII., but this is evidently a mistake, as she was called the *Regent*, while still on the stocks, in the payment warrants. She was blown up in the action with the French fleet off Brest on August 10th, 1512, having been engaged, as the strongest ship in the English fleet, for two hours with the *Cordelière*, the largest ship on the French side, eventually setting her on fire; the French captain, Primauguet, finding it impossible to extinguish the flames, succeeded in running his ship alongside of, and securing her to, the *Regent*, with the result that both ships blew up with all hands.[2] The *Cordelière* is said to have had a crew of 1,200 men, and the *Regent*, which was commanded by Sir Thomas Knyvet, Master of the Horse to the king, is described by Hill as a first-rate, and as the most powerful ship in the Lord Admiral's fleet, the second largest ship being the *Sovereign*, which in the same action was commanded by Sir Charles Brandon. As tending to show that port-holes were still a novelty in the early part of the sixteenth century, it may be as well to point out that a Father Daniell, a priest, who served in the French fleet as a chaplain during the war with France, 1522-24, and who wrote a description of the fighting,[3] off the Isle of Wight on the 18th, 19th, and 20th July, 1523, draws attention to the fact that "the English ships were built with port-holes for their ordnance;" and while he adds that, "the custom was not of long standing," he yet admits that the custom was in use when Louis XII. came to the throne in 1486, which is a further proof, if any is needed, that they had been introduced before Descharges built the

1 Charnock, "History of Marine Architecture."

2 Hill, "Naval History of Britain," 1756.

3 Entick, "A New Naval History," 1757.

Cordelière, his first ship, in 1500. Light will probably be thrown on this and other points connected with warships of that period when the selections from the State Papers in the Record Office and private collections bearing on naval matters in the reigns of Henry VII. and Henry VIII., now being made by the Navy Records Society, are published. It may be mentioned here that the loss of the *Mary Rose*, a vessel of 500 tons, at Spithead, in the early stages of the action of the 18th-20th July, 1523, is attributed by Sir Walter Raleigh to the fact that her lower-deck port-sills were only 18 inches out of the water, and these being open, to enable her to fight her guns, when

FIG. 1.—GUN RECOVERED FROM WRECK OF THE "MARY ROSE;" TEMP. HENRY VIII.

a squall struck and heeled her over, she filled through them and went down.[1] In 1836 many of the guns were recovered from the wreck, with their carriages, among them being a breech-loader of wrought iron of an 8-inch calibre, and others of brass of 5·2-inch bore, now in the museum at Woolwich (fig. 1). She was built in 1509, and had a similar armament to the *Sovereign* after her reconstruction.

After the loss of the *Regent*, Henry VIII. gave directions for the construction of another large ship to take her place; this vessel, known as the *Henry Grace-à-Dieu* was launched at Deptford in 1515, and although authorities differ as to whether her tonnage was 1,500 or only 1,000,

[1] According to Mr. Oppenheim, she was capsized off Brading on July 20th, 1545.

there is no doubt she was the finest ship yet built for the English Navy. In the original MS. in the Pepysian Library her armament is given as follows:

Gonnes of Brasse.

Cannons	4
Demi-Cannons	3
Culverins	4
Demi-Culverins	2
Sakers	4
Cannon-Perers	2
Fawcons	2

Gonne Powder.

	Lasts[1]
Serpentyn Powder in Barrels	2
Corn Powder in Barrels	6

Gonnes of Yron.

Port Pecys	14
Slyngs	4
Demi-Slings	2
Fowlers	8
Baessys	60
Toppe peces	2
Hayle shotte Pecys	40
Hand Gonnes complete	100

Shotte of Yron.

For Cannon	100
,, Demi-Cannon	60
,, Culverins	120
,, Demi-Culveryns	70
,, Sakers	120
,, Faucons	100
,, Slyngs	100
,, Demi-Slings	50
Crosse Barre Shotte	100
Dyce of Iron for Hayle-Shotte	4

Shotte of Stoen and Leade.

For Canon-Perer	60
,, Porte Pecys	300
,, Fowlers	100
,, Toppe Pecys	40
,, Baessys Shotte of Leade	2^m

Then follow other details of stores, among which we find "10́ Coyll of Ropes of Hempe for Wolyng and breching," "Canvas for Cartowches 1 quarter," "Paper Ryal for Cartouches, 6 quarters," etc.

A long list of the different descriptions of guns in use at this period is to be found in the celebrated "Treatise of Artillerie," written by Tartaglia,[2] an Italian, who dedicated his work to King Henry VIII.; in the table referred to he

[1] A Last of powder = twenty-four barrels, about 4,000 lb.

[2] Tartaglia, "Three Books of Colloquies concerning the Art of Shooting."

gives no less than seven different culverins, the iron ball for the lightest being $9\frac{1}{4}$ lb. in weight, and for the heaviest, $79\frac{1}{2}$ lb.; six types of cannon, the iron ball for the lightest, $13\frac{1}{4}$ lb., and for the heaviest, $79\frac{1}{2}$ lb., and so on with other guns. His book gives a great deal of information concerning both the theory and practice of gunnery as then understood, and proves him to have been a man of great talent. He explains the motion of a projectile, taking the resistance of the air into account, the theory of gunpowder, etc., and gives directions for sighting ordnance, and explains the use of the gunner's quadrant, which he invented, and enters into much detail regarding the rules to be observed in the manufacture of ordnance, carriages, etc. As, however, he does not mention what guns were used for sea service, the subjoined table of guns mounted in ships in Queen Elizabeth's reign has been taken from a treatise by Sir William Monson, a distinguished naval commander and writer of that period;[1] a comparison of the two tables showing that practically the same guns were in use in both reigns.

Names of the Pieces of Ordnance	Bore of Cannon	Weight of Cannon	Weight of Shot	Weight of Powder.	Number of Shot in Last of Powder	Point-Blank	Random.
	Inches	Lb	Lb	Lb		Paces.	Paces
Cannon-royal . .	$8\frac{1}{2}$	8,000	66	30	80	—	1,930
Cannon . .	8	6,000	60	27	85	17	2,000
Cannon-Serpentine	7	5,500	$53\frac{1}{2}$	25	96	20	2,000
Bastard-Cannon .	7	4,500	$41\frac{1}{2}$	20	120	18	1,800
Demi-Cannon	$6\frac{3}{4}$	4,000	$33\frac{1}{2}$	18	133	17	1,700
Canon-Petro . .	6	4,000	$24\frac{1}{2}$	14	171	16	1,600
Culverin . . .	$5\frac{1}{2}$	4,500	$17\frac{1}{2}$	12	200	20	2,500
Basilisk . . .	5	4,000	15	10	240	—	—
Demi-Culverin . .	4	3,400	$9\frac{1}{2}$	8	300	20	2,500
Bastard-Culverin .	4	3,000	7	$5\frac{3}{4}$	388	18	1,800
Sacar . . .	$3\frac{1}{2}$	1,400	$5\frac{1}{2}$	$5\frac{1}{2}$	490	17	1,700
Minion . . .	$3\frac{1}{2}$	1,000	4	4	600	16	1,600
Falcon . . .	$2\frac{1}{2}$	660	3	3	800	15	1,500
Falconet	2	500	$1\frac{1}{2}$	$1\frac{1}{4}$	1,950	14	1,400
Serpentine . . .	$1\frac{1}{2}$	400	$\frac{3}{4}$	$\frac{3}{4}$	3,800	13	1,300
Rabinet . . .	1	300	$\frac{1}{2}$	$\frac{1}{2}$	4,800	12	1,000

[1] "Naval Tracts," by Sir William Monson.

No mention is made in the foregoing list of certain guns forming part of the armament of the *Henry Grace-à-Dieu*, such as "Cannon-Perers," "Port-Peecys," "Slyngs," "Fowlers," "Baessys," etc., but in another book, the "Practyse of Artillerie,"[1] published soon after Sir William Monson's treatise, cannon-periors and petrieraes are described as: "Such Peeces as onely shoote stone, or else Murthering shot, both which and Fireballs may be likewise shot out of the aforementioned Ordnance." The length of these guns, as also of the port-peeces, fowlers, slings, etc., was eight or more diameter of the bore, and although cannon-periors were constructed which threw a heavy shot, yet it is almost certain that those carried by the *Henry Grace-à-Dieu* were, like the port peeces, slings, etc., only small swivel guns, throwing balls from 1 to 3 lb. Many of the smaller guns were breech-loaders. In Chapter XIII. of Norton's work, "Port Peeces and Fowlers" are described as, "Brasse cast Peeces open at both ends, inuented to be loaded with Chambers at the Breech-end, fitted close thereinto with a shouldring, euen as the wooden trees for water pypes haue tapred ends to let them close one into another; the shot and wadde being first put into the Chase, then is the Chamber to be firmly wedged into the Tayle of the Chase and Carriage, and being discharged, their Chambers are to bee taken out and fil'd againe and others to be put in ready charged in the place thereof. These Peeces are vsually loaded with $\frac{1}{4}$ or $\frac{1}{3}$ of the weight of their shot in corne powder. Port Peeces and Fowlers are vsually made of cast Brasse, but Bases, Slyngs and Murtherers are commonly of wrought iron; the length of the Base is about 30 times her Calibre, the Sling about 12 times, the

[1] "The Gunner, shewing the Whole Practyse of Artillerie: by Robert Norton, one of his Maiesties Gunners and Engineers." London, 1628.

Murtherers, Port Peeces and Fowlers 8 at the most besides their Chambers, their Chambers about 3 times their Calibres in length and weigh the 6 or 8 part of the whole Chase." These breech-loading guns were known as "Petrieroes a Braga" (fig. 2), and the chamber or breech-block was called the Moscolo, and they were loaded as follows: "The Powder is put into the Moscolo or Chamber and is closely shut up with a Tampion and a Shot put into the Chase or Bore with a Wadd before and behind, and the Moscolo is then fitted into the Braga and closed with a coin of iron behind."[1]

From the description of the guns given above it will be

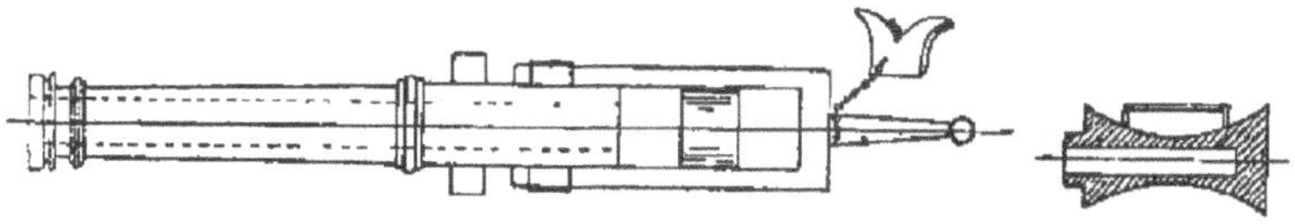

FIG. 2.—PETRIEROE A BRAGA AND THE MOSCOLO.

seen that the armament of the *Henry*, although numerically large, was not heavy, that, in fact, she only carried four 60-pounders, three 32-pounders, four 18-pounders, and two 12-pounders, all the remainder of her guns being 3, 2, and 1-pounders. Many writers of naval history have represented the ship as carrying fourteen heavy guns on the lower deck, and twelve on the main, but this is clearly an error, if the MS. in the Pepysian Library is correct, and there appears to be no reason to doubt its authenticity. Ten of the guns fired through the stern ports, and as many as twenty (if we are to believe old prints of the ship) fired fore and aft through the bulkheads of the poop and forecastle to sweep the waist in the event of an enemy gaining a footing. Other guns, again, were mounted on the broadside of the

[1] "A Treatise of Artillery or Great Ordnance," by Sir Jonas Moore. London, 1683.

poops and forecastles, tier on tier, for at that period those parts of the ship were built like veritable castles.

Although the ordnance remained the same, yet in Queen Elizabeth's reign the ships had much improved, and one result was, that their guns were carried better. Raleigh in his "Discourse on the first Invention of Ships and the several parts thereof," states:—"We carry our ordnance better than we were wont, because our nether overloop [the lowest deck] is raised commonly from the water, to wit, between the lower port and the sea. We have also raised our second decks and given more vent thereby to our ordnance lying on our nether overloop." Sir William Monson, in his "Naval Tracts," to which reference has already been made, draws attention to what must have undoubtedly been an improvement:—"No ship," he says, "now commonly carries greater peeces than a demi-cannon [32-pounder] and the rest of her peeces ought not to be above seven or eight feet long, as they are easier in charging, easeful to the ship, besides better in traversing and mounting."

In Charles I.'s reign in 1637 was launched the *Sovereign of the Seas.* She carried 102 brass guns, and was the first three-decker of which there is any record. According to Mr. Oppenheim she was built for 90 lighter guns, but the king altered the number to 102 and increased their weight, so it is not wonderful that she proved top-heavy at sea. Thomas Heywood, in a description of the ship dedicated by him to the king, thus writes of her:[1] "She has three flush deckes and a forecastle, an halfe-decke, a quarter-decke and a round house. Her lower tyre hath thirty ports, which are to be furnished with demi-cannon and whole cannon throughout, being able to beare them. Her middle tyre hath also thirty ports for demi-culverin and whole culverin.

[1] Charnock, "Marine Architecture," vol. ii.

Her third tyre hath twenty-six ports for other ordnance. Her fore-castle hath twelve ports and her halfe-decke hath fourteen ports. She hath thirteen or fourteen ports more within-board for murdering pieces, besides loop-holes out of the cabins for musket-shot." Had guns been mounted in all her ports she would have carried 126, but it does not appear that she ever had more than 102, so the probability is that the twenty ports right forward and right aft were never filled. The whole cannon she carried were probably the bastard-cannon of Sir W. Monson's list,[1] with a 7-inch calibre and throwing a $41\frac{1}{2}$ lb. shot, the 42-pounder of a little later date, as in an abstract drawn up by command of Charles II. in 1677, the earliest of its kind which appears to be in existence, the heaviest guns carried by first-rates (100-gun ships) were Cannon VII. on the lower decks, and if any further proof were wanted that Cannon VII. was the 42-pounder under a different denomination, it may be found in the fact, that such first-rates in the list of 1677 which survived into the early part of the new century appear in the gun establishment of that time with no other difference in their lower-deck armament than the substitution of the "42-pounder" for the "Cannon VII."[2]

Under Cromwell's government a commencement was made to reduce the number of and re-classify the confused assortment of guns then in use; thus, in the abstract above mentioned, we find that the "cannon-petro" had already changed its name to 24-pounder; and a 12-pounder (the ancient basilisk) also appears there. Similarly, shortly afterwards, the whole culverin and demi-culverin became

[1] Venn in his "Compleat Gunner," published in 1672, terms these guns as "Demi-Cannon" of the greater size or "Cannon of Seven," evidently the "Cannon VII." of a few years later.

[2] James, "Naval History of Great Britain."

the 18 and the 9-pounder, the saker becoming the 6-pounder, although it was not until the close of the century that guns began to be officially designated by the weight of the projectile they fired.

In a memorandum of Sir Cloudesley Shovell to the Admiralty, dated 1690, and in his own handwriting, are the following interesting remarks with regard to the armament of ships:

"Our lower-deck guns are too big and the tackles ill fitted with blocks, which makes them work heavy; the Dutch who have light guns have lignum vitæ sheaves.

"To have ten rounds for our upper tier more than our lower; first, because we can fire the upper guns fastest; next, because we can fight with our upper when we cannot with our lower.

"Eight men should be allowed to an iron cannon and seven to a brass one; seven men to an iron demy-cannon, and six to a brass one; six men to an iron twenty-four pounder, and five to a brass one; five men to an iron eighteen-pounder, the same to a brass one; four men to a twelve-pounder of both sorts; four men to an iron nine-pounder and three to a brass one; three men to a six-pounder or saker.

"The Dutch guns are seldom larger than twenty-four pounders; Admiral Allemond's lower tier are twenty-four pounders brass, his middle, eighteen-pounders brass, his upper, twelve-pounders, his half deck, six-pounders.

"Quick match takes fire of itself except well prepared.

"Cases are wanted for carrying cartridges in time of fight."

It will be seen from this memorandum that the old nomenclature of guns was by this time practically a thing of the past. In Queen Anne's time the establishment for first and second-rates was 32-pounders on the lower deck (42-pounders being sometimes mounted in larger first-rates), 18-pounders on the middle, and 9-pounders on the main and upper decks; for 80-gun ships, 24-pounders on the lower, 12-pounders on the middle, and 9-pounders on the upper deck; for 70 and 60-gun ships, 24-pounders and 18-pounders on the lower deck, and 9-pounders on the upper; for 50-gun ships, 12-pounders below and 6-pounders on the upper decks. Another notable improvement made about this period was in the gun-carriages; in fig. 1 (p. 4), which shows a gun as mounted on the *Mary Rose* in 1523, it will be observed that it lies on a wooden bed for carriage, the rear or inner end only of which is raised off the deck by two small trucks or wheels; this clumsy method of mounting guns seems to have remained in force for over a hundred years, as it was not until the early part of the seventeenth century that two more trucks were added to the fore or outer end of the carriage, and even in 1683 the practice of having four trucks does not appear to have been universally adopted, for Sir Jonas Moore, to whose "Treatise on Artillery" reference has already been made, thus writes: "At sea the ordnance are mounted upon small carriages, and upon four, and sometimes only two, low wheels without any iron work. 'Petrieroes a Braga,' 'Harquebuss a Croc,' etc., are mounted on strong pins of iron having rings, in which are placed the trunnions with a socket, so that they are easily turned to any quarter." Crude, however, as the gun-carriages were, probably, at that period, yet they were already being constructed on the same pattern, which, although with improvements, practically obtained in the wooden truck-carriages down to our own time, even the names given to the different parts having been handed down unaltered; if we refer again to Sir Jonas Moore, we find him thus defining the "Qualifications of him who takes the charge of the cannon of a ship."

"The Gunner ought to be a man who knows the Goodness of a Peece of ordnance, the force of Powder, and who also knows how to mount a Peece of Ordnance upon its carriage, and to furnish it with bolts, Plates, Hooks, Cap-squares, Axletrees and Trucks; to order well its Breeching and Tackles; to plant the Cannon to purpose in the middle of the port; to know how to make ready his cartridges; to see that the beds and Quoins be firm and in good order; to see that his Companions renew their priming Powder every evening; that the Linstocks be ready and furnished with match; that 3 or 400 Cartridges are ready filled," etc.

The method of loading has been practically always the same, with the exception of the early breech-loading guns already described; sponges, rammers, and ladles being supplied under those names from the first, but there was a difference, and an important difference, in loading between the guns for sea and land service, and it lay in the use of the ladle for the powder. "The Ladle is put into the barrel, and when filled is pushed gently into the bore, quite home to the touch-hole, it is then turned round to free the powder, the Ladle withdrawn, and the Powder pressed softly home with the Rammer, after which a good Wadde is thrust home to the powder with three or four hard strokes and then the shot."[1] It is evident that this method of loading was eminently unfitted for ship work, and cartridges for sea service were introduced at a very early date, as is shown by the list of stores quoted in the earlier part of this chapter as supplied to the *Henry Grace-à-Dieu,* where canvas and paper for "cartouches" figure among the items; the use of cartridges for naval guns is also confirmed from other sources; thus in Sir Jonas Moore's "Treatise" we find: "At sea to load Great Ordnance they never use the Ladle but Cartridges of Canvass

[1] "The Complete Cannoniere, or the Gunner's Guide," by John Roberts. London, 1639.

or Parchment as well for expedition as secnrity, to which they give three Calibres in breadth and four in length and two halfe-Calibres for the lying in case there is no Scrole, they open that end towards the Touch-hole with a knife or Priming Iron upon that part which corresponds to the Touch-hole;" so also in his memorandum Sir Cloudesley Shovell emphasizes the necessity of having cases for carrying the cartridges.

To enable the gun to be fired, it was primed, that is, powder was poured down the touch-hole or vent by means of a priming-horn, and a slow match was then applied. Slow match was probably introduced at a very early date, and remained in use until the end of the eighteenth or beginning of the present century; it was supplied as the "match and staff,"[1] the staff being about two and a half feet in length, and the match being wrapped round it, the end being secured by being thrust through a mortice at the extremity and fastened by a peg. A tub was also provided in which to keep the slow match when not in use.

At the beginning of the present century the largest guns in common use were the long 32-pounders and 42-pounders, but a much shorter and lighter gun, called a carronade, was also used for ships' armaments. The carronade was so-called from having been introduced by the Director of the Carron Foundry, in Stirlingshire, where in 1776 "carronades" or "smashers" were commenced to be made, and it was first brought into the service in 1779. Its peculiar advantage consisted in the capability of projecting shot of large calibre with accuracy to such distances as vessels of war were supposed to engage at, viz., from 400 to 600 yards, with a great saving of metal, powder, and in the number of men required to work it; thus the 32-pounder carronade, while shorter

[1] Lloyd and Hadcock, "Artillery, its Progress and Present Position."

than the 4-pounder, was lighter than a 12-pounder. It had no trunnions, but was cast with a loop underneath, a bolt passing through which attached them to their carriages (fig. 3). Carronades were constructed of all calibres, from a 6 to a 68-pounder, and nearly all classes of vessels carried a certain proportion of this type of gun. To show the great advance that had been made by the beginning of the present century in the simplification of the armaments of ships of war, that of Nelson's flagship, the *Victory*, is subjoined:

FIG. 3.—32-POUNDER CARRONADE.

lower deck, thirty 42 and 32-pounders; middle deck, thirty 24-pounders; main-deck, thirty 12-pounders; upper dock, ten 12-pounders and two 68-pounder carronades. Throughout the French War the 42-pounder formed the principal armament for the lower decks of the largest of our first-rates, and the regulation to that effect continued in force till 1839, when the 50-cwt. 32-pounder was substituted for it. During the next ten years various experiments were carried out with guns of heavier calibre, principally 42-pounders, but it was found that from their weight they required such large crews to work them as to seriously overcrowd the gun-deck; while in range and penetration they were in no way superior to the latest type of 32-pounder, a long gun of 56 cwt. One heavier solid-shot gun was however introduced at this time, and came into general use as a pivot gun for chase purposes, viz., the 95-cwt. 68-pounder constructed by Colonel Dundas, which proved a most valuable weapon. In the meantime the development of a new type of gun had been going forward which before very long was to sound the death-knell of wooden ships; this was the type of large-calibre guns for the projection of shells and hollow shot, first introduced into the French Navy by Colonel Paixhans in 1824.

Shells, as far as is known, were first used in naval warfare in the latter half of the eighteenth century, but bombs (iron shells filled with gunpowder) are said to have been invented at Venlo in Holland in 1495, and used by the Turks at the siege of Rhodes in 1523. They came into general use about 1634, having been previously used only by the Dutch and Spaniards. Voltaire states that bomb-vessels were built in France in 1681. It is well authenticated that the destruction of the Turkish fleet by the Russians in the action off the mouth of the Liman in the Sea of Azov, on June 17th, 1788, was entirely due to their use. Sir S. Bentham, the distinguished naval architect and engineer was at that time in the Russian service; he mounted a number of long 36-pounders and 48-pounder howitzers on non-recoil carriages of his own invention in barges and small vessels, and with these the Russians made their attack on the Turkish squadron, shelling them at a range at which the Turkish guns were apparently useless. Ten ships of the line were set on fire and blown up, another was sunk, and of the eleven crews, numbering some 10,000 men, only 3,000 were saved. Shells and other incendiary projectiles were supplied to French ships during the Revolutionary and Napoleonic wars, but they seem to have been almost as dangerous to their own vessels, which carried them, as to an enemy; for, exclusive of numerous comparatively trifling explosions, "qui s'est trop souvent renouvelé dans la longue et funeste guerre de la Révolution,"[1] four or five ships of the line, six frigates, and several smaller vessels were either burnt, blown up, or so terribly damaged by their own incendiary and combustible weapons, as to be incapable of further resistance, without injuring or destroying one of ours; as, for instance, in the action of July 13th, 1795,[2] the *Alcide*, seventy-four, was set on fire by her own grenades; of the 615 men on board, 300 were saved by the boats of the British ships; and the setting on fire of the *Orient* at the battle of the Nile has always been attributed to the ignition of some carcases or inflammable composition she had on board. A wholesome dread of all incendiary projectiles, as being more dangerous to the ships carrying them than to an enemy, seems to have restrained our Admiralty at that time from issuing shells to our own ships, except in a few special cases, where shells fitted with time fuses were issued for use against bodies of the enemy's troops; even the use of hand-grenades from the tops was strongly disapproved of; certain it is, that there seems to be no instance on record of an English ship of war blowing up in action during that long-protracted struggle. Shells were used for the purpose of bombarding towns, etc., but they were fired from bombs or mortar-vessels specially fitted for the purpose. On the adoption of Colonel Paixhans' "Canon-obusier" in the French Navy in 1824, however, our authorities decided it was necessary to introduce some analogous weapon into our own service, and after a long series of experiments, the 65-cwt. 8-inch shell gun was

[1] De la Gravière, "Guerres Maritimes," vol. i., p. 97.

[2] Sir Howard Douglas, "Naval Gunnery."

finally adopted in 1838, and at the outbreak of the Russian war had taken the place of the 32-pounder as the lower-deck armament of line-of-battle ships and the main-deck armament of frigates, the 32-pounder, being mounted on the other decks. The length of the 8-inch gun was nine feet, and it threw a hollow shot of 56 lb. in weight, the highest charge being 10 lb. of powder. In 1860, when the days of wooden ships were drawing to a close, the armament of the *Marlborough*, a three-decker of 121 guns, at that time the flagship in the Mediterranean, was as follows :—lower deck, thirty-two 8-inch 65-cwt. guns; middle deck, thirty 8-inch guns; main-deck, thirty-two 32-pounders, 56 cwt. and 9 feet 6 inches long; upper deck, twenty-six 32-pounders and one 68-pounder pivot gun on the forecastle. The armament of the 91-gun line-of-battle ships was :—lower deck, thirty-four 8-inch guns; main-deck, thirty-six 32-pounders; upper deck, twenty 32-pounders, with one 68-pounder pivot gun forward. The 51-gun frigates carried :—main-deck, thirty 8-inch guns; upper deck, twenty 32-pounders with one 68-pounder pivot gun. The 21-gun corvettes carried 8-inch guns, with a 68-pounder as pivot gun; while the smaller corvettes and sloops carried short 32-pounders of 32 cwt. on the broadside, and a long 56-cwt. 32-pounder as a pivot gun. Shortly after the Russian War a larger shell gun, the 10-inch 85-cwt. gun was introduced and mounted on the main-decks of five heavy frigates of a new type, the *Mersey*, *Ariadne*, *Galatea*, *Diadem*, and *Doris;* but with these exceptions the armament as given above was the armament of the British fleet of only thirty-five years ago; in fact, even as late as 1870 the 8-inch shell gun had not entirely disappeared from some of our ships.

Before concluding this chapter, we must retrace our steps somewhat, in order to point out other improvements in the art of gunnery which were effected at the commencement of and during the first half of the century. It has already been mentioned that a match was originally used for firing the guns, which were primed by means of a priming-horn, and that this antiquated method of firing obtained up to the end of last century; a slight improvement had certainly latterly been made by the invention of tubes, or quills, filled with powder, and having a head formed on them by splitting the quills for a short length and spreading the split portion out. They formed a train down the vent, and thus greatly increased the certainty of fire and the rapidity of its communication to the charge; but still the method was complicated. It was to Captain Sir Charles Douglas that the first great improvement in the method of firing was due, when as captain of the *Duke* (1778-1781), a ship of ninety-eight guns, he invented, and supplied to that ship at his own cost, flint-locks,[1] which had the advantage of giving greater safety, and placing the means of firing directly under the control of the captain of the gun. In November, 1781, Sir C. Douglas was appointed to the *Formidable* as Captain of the Fleet, in which ship he likewise introduced locks and other improvements; he was succeeded in the command of the *Duke* by Captain (afterwards Lord) Gardner; and in Rodney's great action of April 12th, 1782, the quick and efficient firing of both the *Duke* and the *Formidable* was so conspicuous that all further opposition to the introduction of locks was borne down. But that battle having likewise put an end to the maritime part of the war, no measures for the supply of locks to naval ordnance appear to have been taken till 1790, when "brass locks of a new pattern were provided and used throughout the great war, and up to the year 1818, when a double-flinted lock, invented by Sir Howard

[1] The Admiralty having refused to approve the innovation.

Douglas, was ordered for general use in the Navy. In 1807 a composition, which could be ignited by friction or concussion, was invented, and as it did away with the necessity for a priming of loose powder, it was one of the most important improvements which had as yet been introduced, although it was not brought into use until many years later. The composition was composed of chlorate of potash, antimony, mealed powder, and sulphur, and, later on, fulminate of mercury was added. The locks were fixed to the vent field on the right side of the gun, so that the hammer would fall clear of the vent, or otherwise it would have been thrown back with great violence by the rush of gas caused by the ignition of the charge and have been injured; but this necessitated the making the tube in a rectangular form, one end of which went down the vent, and the other, filled with composition, lay along the exterior of the gun a couple of inches, so that the hammer could strike and ignite it without going directly over the vent. This construction was found, however, to be so sluggish as not to accomplish the great desideratum in naval gunnery, which is, that the firing of the charge and the shot leaving the gun shall take place as quickly as possible after pulling the trigger line, in order that there may be little time for any alteration to take place, from the motion of the ship, in the aim of the gun. Thus it was necessary to devise some means by which the hammer, after having struck fairly upon the head of the tube placed in the vent, should instantaneously slip, or be drawn aside, so as to be out of the way of the explosion. Various methods were devised, but the most efficient and simple one was that invented by an American named Hidden, and patented in 1842, which, with some modifications and improvements of the hammer made by Colonel Dundas, was introduced into the service a few months later. The hammer, A (fig. 4), which was of wrought-iron of good quality, was fixed into a block, or joint, of gun-metal by means of a pin, *a*, and a slot made to admit of a back motion. The hammer fell by its own gravity upon the vent on the lanyard, B C, being pulled, when, being liberated from the larger part of the slot in which it is made to turn, it instantly shifted its position by the continued action of the lanyard, which pulled it from its first position directly over the vent, as shown by the black lines at A′ in the figure. The weight of the hammer was $3\frac{5}{16}$ lb.[1] The lock and cross-headed quill tube were superseded about the

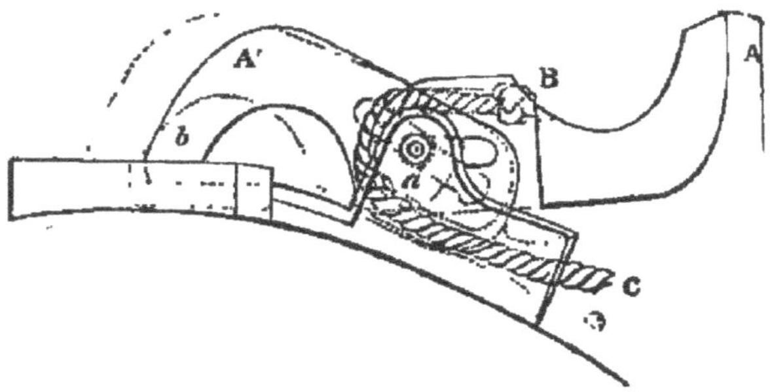

FIG. 4.

year 1862 by the "friction-tube" invented by Captain Boxer, R.A., but a description of this will be given further on.

Mention has already been made that to Sir C. Douglas, in 1778, the invention of locks was due, but another great improvement was introduced by him, and sanctioned by the Admiralty, and that was the substitution of flannel for paper for cartridges. In the early days of naval gunnery canvas was generally used for cartridges, but it was superseded by paper, or parchment, which, however, necessitated the guns being wormed after every discharge, on account of the lower end of a paper cartridge remaining generally at the bottom of the bore in a state of ignition. Flannel remained in use for cartridges up to a late period, but it

[1] "Treatise on Naval Gunnery" (Sir Howard Douglas).

has been replaced by serge and "Amiantive" or silk cloth, and this latter is now almost exclusively used, and is made of refuse silk from outside of cocoons.

Up to the early part of this century the method of laying guns was of the most rough-and-ready description; sights were unknown, and whether a shot struck the mark, unless the range was very close, was very much a matter of chance. The method of aiming consisted in running the eye along an imaginary line on the exterior of the gun, parallel to the centre of the bore, until it reached the object, and, as a gun is smaller at the muzzle than at the breech, allowance had to be made for the inclination of that line to the axis of the bore. The necessity for having a fixed sight (called a dispart) on the muzzle, or centre, of the gun, by which the difference between the thickness of the gun at the breech and muzzle was corrected, seems to have been recognized at a very early period, and we find the method of disparting a gun, as it was termed, described by several writers on artillery in the seventeenth century thus:—"To dispart a Peece is to bring the mettall at muzle equall with the metall at the base ring. But for nicenesse I would make another Dispart upon the Trunnions, and I could wish that the Dispart should be skrewed into the muzle ring of the Peece; also I could advise the Sea-Gunner upon some occasions to use Disparts between the Trunnions of their Peeces." [1]

Sir Jonas Moore in his treatise, to which reference has been already made writes:—"The Dispart Line of the Peece is the difference betwixt the semi-diameter of the Muzzle-Ring and Base-Ring, or a line drawn from the utmost top of the Base-Ring parallel to the Chase of the Peece, the Dispart being the nearest distance betwixt the same line and the top of the Muzzle-Ring." Disparts, how-

[1] "The Complete Cannoniere, or the Gunner's Guide," by John Roberts. London, 1639.

ever, like other sights, did not come into general use until the beginning of the century: a proposal to use sights was sent to Lord Nelson in 1801, bnt the reply was unfavonrable, his lordship expressing a hope "that our ships would be able, as usual, to get so close to our enemies that our shot cannot miss the object." The only means provided at that time for pointing naval ordnauce, were the quarter-sight scales engraved on the sides of the base-rings, in quarter degrees from point-blank[1] to two or three degrees of elevation. For close, horizontal fire, guns could be laid by the point-blank quarter-sight, with sufficient accuracy, by simply bringing the notches upon the base-ring and the swell at side of muzzle to bear upon the object aimed at; the elevation was correct, and, in close action, the line was sufficiently true. It was different, however, when elevation was required, because, unless the line was correctly taken over the top of the gun, great errors in horizontal divergence were produced. It was not until after the conclusion of the war with France that the subject of sights and elevation scales was taken into serious consideration, and various instruments were then devised to give the line of direction and the necessary elevation to the gun at the same time, but many years elapsed before the tangent sight, as it existed at the time when smooth-bore ordnance became obsolete, some thirty years ago, came into use. The brass tangent sight was of hexagonal form, and on the different faces, except one, which had the degrees marked on it, were the ranges in yards corresponding to different charges and projectiles,[2] thus:

[1] Point-blank was the spot where the shot would strike the water if fired from a gun which was laid perfectly horizontal, generally a distance of about 150 yards from the muzzle.

[2] There were three charges used for smooth-bore guns at that time—reduced, full, and distant.

°		for degrees.	
F	,,	yds. Full	Charges for shot.
D	,,	yds. Distant	Charges for shot.
R	,,	yds. Reduced	Charges for shot.
S F	,,	yds. Full	Charges for shell.
S R	,,	yds. Reduced	Charges for shell.

It fitted into a block of gun-metal, which was screwed on behind the base-ring, and it could be used up to five degrees of elevation. The dispart sight was of gun-metal, with its top filed to a sharp edge, and was placed in front of the second reinforce (as it was called), about in line with the trunnions, its height being equal to the dispart at that part of the gun. When a greater elevation than five degrees was required, a wooden tangent scale was used, by which an elevation of ten or twelve degrees could be given; the scale was placed and held vertically on a step of the carriage and used in conjunction with the pendulum, which registered the heel of the vessel.

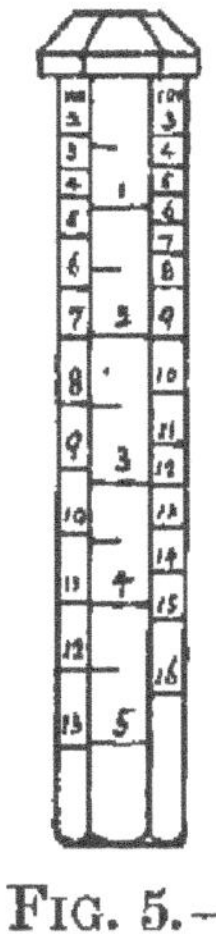

Fig. 5.—Brass Tangent Sight.

Before concluding this chapter it may be as well to give a short description of the truck-carriage, as it was called, which after being in use, with scarcely any changes, since the seventeenth century, has only finally disappeared from the Navy during the last few years, and also to explain briefly the method of working the guns so mounted. The carriage was formed by two wooden sides, termed the brackets, which were joined together by front and rear pieces, called the transoms; there were four wooden wheels or trucks, kept in their place on the axles by iron linch-pins. The trunnions of the guns fitted into bearings hollowed out on the fore upper part of the brackets, and were secured by the cap-squares, iron straps working on a

Fig 6.—32-Pounder Gun, Working with Reduced Crew of Nine Men. "Elevate."

hinge, so that they could be thrown back when necessary for dismounting the gun, etc. The upper part of the inner half of the bracket was cut into three steps, on one of which the handspikes rested when the gun was being elevated (fig. 6). Between the two brackets, and resting on the transoms, was the bed, which was movable, and on this again was placed the large quoin, a wedge-shaped block of wood, which was moved along the bed as requisite to keep the gun at the required elevation after it had been laid by the captain of the gun (fig. 6); the quoins were also marked in degrees, as shown in fig. 7, so that when all

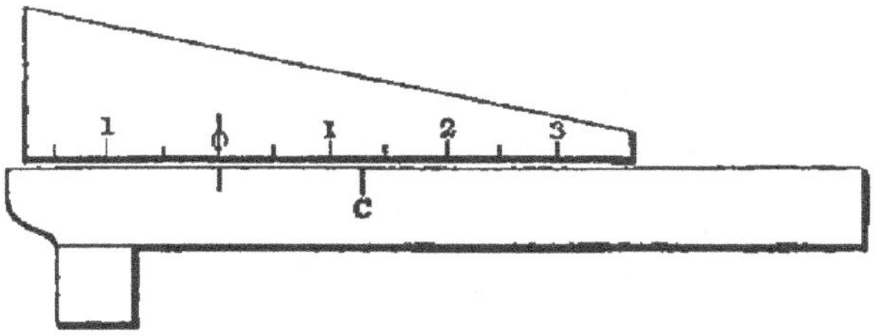

FIG. 7.—MARKED QUOIN AND BED.

the guns were fired together at the same object, the guns were laid for elevation by the degrees on the quoin, the amount of the elevation being ordered by the officer in charge. In training the guns to the right or left on an object, the handspikes were used as levers under the horns (the inner ends) of the brackets, to heave the gun over in the direction required, the men on the side-tackles, which were hooked from bolts on the horns of the brackets to the ship's side, assisting by hauling on them at the same time. There were no means of controlling the recoil, which was merely limited by a stout rope, called the breeching, passed through an eye on the breech of the gun, the ends being secured to ringbolts in the ship's side. When the gun was fired, it recoiled to the full length of the breeching, and was then in the position for loading, the muzzle being just inside the port, and was prevented from running out by a tackle at the rear, called the train-tackle (latterly the preventer-tackle); when loaded, the gun was run out again by the side-tackles.

In stationing the men at the guns, it was the custom to give a full gun's crew to only the guns one side of a ship's battery; when it was required to fight both sides at the same time, half the crew went over and manned the opposite guns. The normal crew for a 56-cwt. 32-pounder was 14 men, and the men were told off and stationed in order as follows:—No. 1, the captain, in rear of the gun, facing the port; No. 4, the sponger, to the right of gun, close to the ship's side; No. 3, the loader, to the left of the gun, close to the ship's side; No. 6, the assistant-sponger, to the right of the gun, next to No. 4; No. 5, the assistant-loader, to the left of the gun, next to No. 3; No. 2, the second captain, on the right of No. 1, facing the ship's side, clear of the recoil. These six numbers were called the gun-numbers, and the numbers above them were called the auxiliaries, who were stationed in the same way, Nos. 7, 9, 11, 13 in order on the left of the gun, and Nos. 8, 10, 12, 14 on the right; the handspikemen were Nos. 9 and 10 (fig. 8). When the guns were cleared for action, each gun's crew cleared away two guns, their own gun and the gun on their right, and the order was then given to man the starboard or the port guns as desired. The captain of the gun was

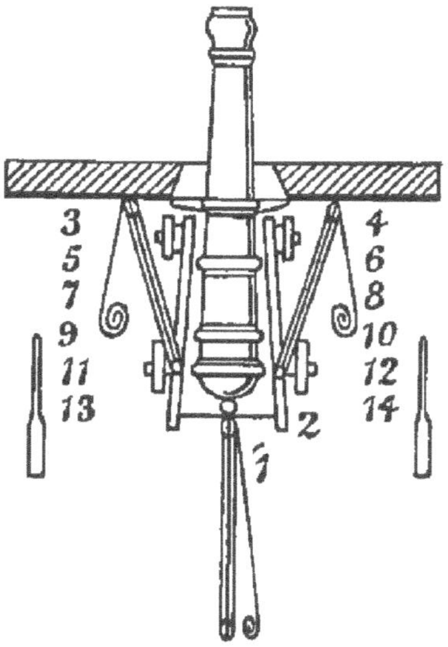

FIG. 8.—32-POUNDER GUN'S CREW, CLOSE UP.

responsible for the proper carrying out of their duties by the other numbers, and that the necessary stores were provided; he was further responsible for the proper laying and training of the gun. After each round the gun was sponged by No. 4, who then rammed home the cartridge, shot, and wads, the sponge and rammer being handed to him by No. 6, who also replaced them on the deck after they had been used. The cartridge and shot were entered by No. 3, the powder-boy bringing the cartridge case close up to the gun for the purpose, and No. 5 handing the shot, No. 3 also assisted No. 4 in ramming home. The most important duties of No. 2 were attending the quoin during the elevation of the gun, putting the tube in and cocking the lock at the order "ready," or half-cocking again, if the order was given.

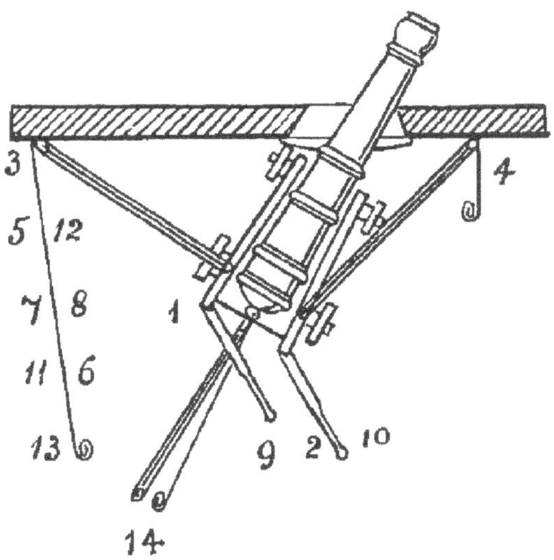

FIG. 9.—32-POUNDER GUN'S CREW. EXTREME TRAIN "RIGHT"

When laying the gun, No. 1 retired to the full extent of the trigger line, bending sufficiently to get his eye along the sights, leaning well over on his right knee, but keeping his left foot clear of the recoil, so that he could spring safely on one side as soon as the gun was fired. Similarly, No. 2 was carefully taught to step up clear of the gun when cocking or uncocking, so as to avoid injury in the event of a premature explosion of the charge. In training the gun on the object, No. 1 gave the order "muzzle right" or "left," as requisite, the handspikes and side-tackles being worked accordingly. In laying the gun for elevation, No. 1 gave the order "elevate," the handspikemen then placing their handspikes on the step of the carriage, and under the gun, raised or lowered it as No. 1 moved his hand up or down (fig. 6). As soon as he was satisfied, No. 1 gave the order "well." No. 2 then forced in the quoin, and as soon as he felt the weight of the gun, gave the order "down" to the handspikemen, springing up to the safety position on the right, when the next order would then be "ready." If it was necessary to give the gun extreme training, the right or left side-tackle was shifted and the men worked as shown in the figures 9 and 10.

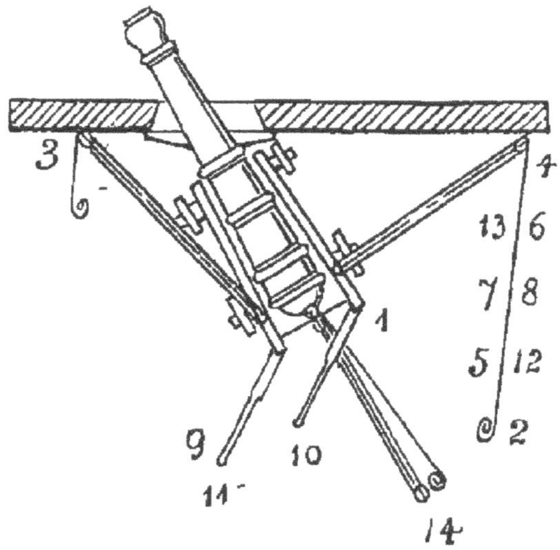

FIG 10—32-POUNDER GUN'S CREW. EXTREME TRAIN "LEFT."

There were practically only two methods of firing practised: "independent firing," when the captain of each gun laid and fired his gun independently, and "broadside or concentrated firing," in which the guns were laid for the same object, the bearing of the object, heel of ship, and distance being given from the upper deck; the guns were then laid and trained according to these directions, and the guns fired simultaneously by order. The necessary elevation was given to each gun either by means of the degrees marked on the quoins (fig. 7), or by a wood scale held vertically with one end on deck, so that a mark on the rear of the gun could be brought into line with the required degree marked on the graduated scale. The training was obtained by means of what were called converging lines; the beams over each gun were marked to denote, "right abeam," one point and a half before or abaft the beam, and three points (extreme training) before and abaft the beam. Each converging line had one end hooked exactly over the centre of the port, the other end being held by No. 1 (in rear of the gun) immediately under the required mark on the beam overhead, and the gun was then trained until its axis corresponded with the converging lines.

Originally broadsides were only fired by a "directing gun," the gun whose captain was considered to be the most reliable marksman being selected for that purpose, the No. 1 giving the elevation and moment of firing to all the others, the bearing on which the broadside was to be delivered being usually decided from the upper deck.

About 1850 the late Captain Moorsom, R.N., an officer of high scientific attainments, introduced an instrument called the "Director," which in these later days has been brought to great perfection, and by means of which the bearing, heel, and distance were all determined, and the guns laid accordingly; it was generally fixed on the upper deck, over the centre main-deck gun, or other convenient position, and the signal or order to fire was given by the officer attending it as soon as he found the object coming on with its sights.

The method described above of firing "converging" or "concentrated broadsides" was still in use on board the frigates forming what was called the "Detached Squadron" not more than twenty years ago.

A description of the various projectiles in use must be deferred to another chapter devoted to that branch of our subject.

CHAPTER II.

THE FIRST INTRODUCTION OF RIFLED GUNS—THE ARMSTRONG BREECH-LOADING AND OTHER SYSTEMS.

It is no exaggeration to say that since the Russian War gunnery has undergone a complete revolution, and greater advances have been made during the last thirty-five years, than during the whole long period which has elapsed since the first introduction of gunpowder more than five hundred years ago.

As early as 1545 cast-iron ordnance were manufactured in England, although bronze guns, and guns made of iron bars hooped together, had been constructed for nearly two hundred years previous to that date, but for some three hundred years, or rather more, the cast-iron process of manufacture only was employed. After the introduction in 1856 of Mr., now the present Lord, Armstrong's method of building up guns, wrought-iron, in conjunction with steel for the inner tube, superseded cast-iron, but since 1880 steel has in its turn replaced all other materials for constructing guns; moreover, during the last thirty-five years we have gone from the 68-pounder smooth-bore gun, 10 feet long, and with an extreme range of barely 4,000 yards, to the 111-ton rifled breech-loading gun, 43 feet 8 inches long, throwing a projectile 1,800 lb. weight to a distance of 12,000 yards.

The practice of loading guns at the breech is not of modern date; for, as was mentioned in the previous chapter, many descriptions of breech-loading cannon were in use at the beginning of the sixteenth century, one or two specimens of which were recovered from the wreck of the *Mary Rose*, sunk at Spithead in the action with the French on July 18th, 1523. These guns were made of wrought-iron bars, secured with iron hoops and fixed in a solid bed of elm, 9 feet 8 inches long; they were loaded at the breech by a detached chamber, which was kept in its place by a chock of elm. Several ancient chambers of guns, of smaller sizes, of wrought iron, of the time of Henry VIII., have been found at Dover. A small brass 4-pounder, with a detached chamber for loading at the breech, and bearing the cypher of the Dutch East India Company, was found on an islet on the coast of Australia, upon which a Dutch ship, named *Zeevyk*, was wrecked in 1727, and in the Museum of the Royal United Service Institution is an old Indian bronze breech-loader, the detached powder-chamber being kept in place by a key or wedge.

In 1742 a Mr. Benjamin Robins published one of the most important works on gunnery ever written, and which has been commented upon by nearly all later writers on this subject. He invented the "ballistic pendulum," the first instrument of any accuracy for measuring the velocities of projectiles. He explained the value of rifling, and the advantages to be gained by the use of elongated projectiles, having mastered to a great extent the theories of the resistance of the air. In his "New Principles of Gunnery" he made the following prophecy:—"Whatever State shall thoroughly comprehend the nature and advantages of rifled barrel pieces, and, having facilitated and completed their construction, shall introduce into their armies their general use, with a dexterity in the management of them; they will by this means procure a superiority which will almost equal anything that has been done at any time by the particular excellence of any one kind of arms."[1] Two hundred years, however, before Robins wrote, experiments must have been carried out with "rifled" barrels, for there is at Woolwich a barrel, with the date 1542 on it, rifled with six fine grooves, having a twist of one turn in 26 inches.

It was not till a hundred years after Robins published his work that any attempt was made seriously to rifle guns, but in 1846 a Major Cavalli, a Sardinian artillery officer, and Baron Wahrendorff, a Swedish officer, both invented breech-loading rifled iron guns, with which experiments were carried out at Shoeburyness and in Sweden, cylindro-conical and cylindro-ogival-headed shot being used.

The length of the Cavalli gun was 8 feet 10·3 inches; it weighed 66 cwt., and had a calibre of $6\frac{1}{2}$ inches. Two grooves were cut spirally along the bore, each of them making about half a turn in the length, which was 6 feet 9 inches. The chamber, which was cylindrical, was 11·8 inches long, and 7·0008 inches diameter. The breech was closed by means of a wrought-iron case-hardened wedge, which, when in its place, covered the extremity of the chamber; the projectile being entered, and the cartridge placed behind it, a "culot," or false breech of cast iron was made to enter $2\frac{1}{2}$ inches into the bottom of the chamber behind the cartridge, and the wedge before-mentioned, which worked in a slot in the side of the gun, was then pushed home until it completely covered the chamber. The projectiles, which weighed 66 lb. when hollow, and 100 lb. when solid, were $16\frac{3}{4}$ inches in length, with a diameter 6·4 inches; there were two projections, a $\frac{1}{4}$ inch deep, directly opposite to each other, running along the cylindrical portion of the shot, which entered the grooves

[1] Page 312, chap. xiii.

in the rifled bore; these projections made an angle of 7° 8′ with the axis of the shot. There was a double percussion agent for the shells; one part an ordinary percussion cap, and the other consisting of four small capsules of glass containing sulphuric acid, placed about the neck of the shell, the intervals being filled with chlorate of potash, while the upper part of the neck was filled with common gunpowder and stopped with wax.[1] A further series of trials were carried out in Piedmont in 1853 and 1854, when, with an elevation of 10 degrees, a mean range of 3,058 yards was obtained, with a mean deviation to the right of 3·4 yards, and to the left of 3·3 yards; during the final trials, with an elevation of 25 degrees, the mean range was 5,563 yards, with a mean deviation of 3 yards to the right and 4 to the left. The rifled gun constructed by Baron Wahrendorff differed but little from that of Major Cavalli; while he used a wedge similar to that in the Cavalli gun, the other arrangements were somewhat more elaborate. Although good results were obtained from both guns, yet they were never perfected, and seem soon to have been dropped.

A Mr. Lancaster was the next to enter the field; his idea was to cause the shot to rotate on its axis by giving the gun an elliptical bore of small eccentricity. The spiral, which had a turn of about one-fourth of the periphery in its length, was not uniform, being nearly rectilinear to a certain distance from the bottom of the bore, and increasing gradually, but rapidly, in curvature from thence to the muzzle. Several guns were rifled on his principle, and were sent out to the Crimea and mounted in some of the gunboats employed in the Baltic and Black Sea, but unfavourable reports were made of these guns, both at Bomar-

[1] Sir H. Douglas, "Treatise on Naval Gunnery."

sund and in the Crimea; the shot sometimes jammed in the bore, some of the guns burst, and the flight of the projectile was often erratic, so they were withdrawn in favour of the 68-pounder; still, as was proved in the experiments with the Cavalli gun, it was shown that a greater range could be obtained with elongated projectiles than was possible with spherical.

It was now that the man, whose name has become world-renowned, and will remain inseparably connected with the revolution effected in modern gunnery, first appeared on the scene. In the latter part of the year 1854, Mr. William Armstrong (now Lord Armstrong) submitted to the Duke

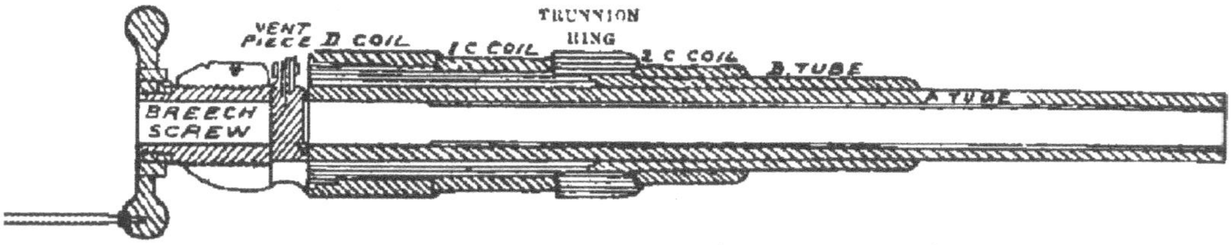

FIG. 11.—40-POUNDER ARMSTRONG GUN.

of Newcastle, then minister of war, a proposal for a rifled field-piece on a new principle, and undertook to construct a gun upon the plan he had suggested. Two years passed before the gun, completely equipped, was submitted for trial; it then became the subject of a long course of experiments, which ultimately led to the general introduction of six different natures of his guns into the service, viz., the 7-inch, 40, 20, 12, 9, and 6-pounders.

The principle he adopted in construction was to build up the gun of separate parts (fig. 11), and this principle was carried out in all his guns, small or large. The constituent parts were as follows: the A-tube or barrel; the breech-piece; three to six different coils or jackets, according to size of the gun; the trunnion-ring; vent-piece; tappet and lever rings; and with the 7-inch and 40-pounders an indicator ring, which showed when the breech-screw had screwed the vent-piece quite home. In the "A-tube" were the bore, shot chamber, and the powder chamber; the bore was rifled on the polygroove system (*i.e.*, with a large number of shallow grooves), having a uniform twist (fig. 12); at the end of the powder chamber in all guns, except in the 7-inch, was screwed a copper ring, against the outer edge of which the copper facing of the vent-piece fitted, and so closed the bottom of the bore. In the 7-inch guns the ring in the powder chamber was of wrought iron. The barrel of Lord Armstrong's first gun was of forged steel, but it being found extremely difficult to get flawless ingots of steel for boring out the barrels of even his smaller guns at that time, wrought iron was substituted and used for the purpose for some years. The "breech-piece" was a solid forging of wrought iron; it was bored, turned, and shrunk on to one end of the barrel. The breech-piece of the larger guns was welded to the coil in front of it. A slot or opening was drilled through it and the coil above, to admit the vent-piece, and behind the slot a screw was cut in the interior surface of the breech-piece, into which the breech-screw fitted.

FIG. 12.

It may be seen from the drawing (fig. 11) that the vent-piece was kept against the end of the barrel by the breech-screw, which was supported by the breech-piece, and the latter had therefore to withstand a longitudinal strain, viz., that exerted by the gas against the vent-piece tending to force it backwards, and it was for that reason that the breech-piece was made of a solid forging with the fibre of the metal disposed so as to resist longitudinal strain.

The "trunnion ring" was also a solid wrought-iron forging, which was turned, bored, and shrunk on to the gun, the fibre of the metal being in the contrary direction to that of the breech-piece. The "jackets" or "coils" were made by winding a long bar when hot round a mandrel, and so forming a cylinder, which was then welded, turned, and bored to the required dimensions for passing over either the barrel or the first layer of wrought-iron jackets; the fibre of the metal thus running circumferentially, presented the best condition for resisting radial strains.

The gun was built up by passing the cylinders, or jackets, previously expanded by heat, over the barrel, and then letting them cool. As the internal diameter of the jacket was less than the external diameter of the barrel, when both were cold, it is evident that this method of building up compressed the barrel and expanded the jacket, by which a great increase of strength was gained, as every particle of the metal was made to take some share of the strain caused by the internal pressure resulting from the explosion of the charge. The great advantage arising from this method of building up guns, instead of casting them in one piece, is that it minimizes the chances of flaws, as with every casting there is a possibility, no matter how much care is exercised, of having a cavity, or flaw, in the metal, which, undiscovered, would be a permanent source of danger. After the gun was built up it was brought to the necessary dimensions by turning and boring. It was then rifled, the number of grooves depending on the size of the bore, but they were all of the same character, though not exactly of the same size; the form is shown in fig. 12, the driving side being radial to the bore. The lands (the spaces between the grooves in the bore of a gun) were left narrow so as to remove as little as possible of the lead with which the projectiles were coated to take the grooves of the rifling. After being rifled, the gun was prepared for the breech mechanism by having a slot cut in the upper part of the breech for insertion of the vent-piece, and underneath this a circular hole was drilled through the lower side of the gun, which was called the "water escape." The gun was then chambered, *i.e.*, the powder chamber and shot chamber were bored out to their respective dimensions. The diameter of the former was rather the greater, removing all the lands of the rifling; in the latter, which was slightly conical, the whole of the lands were not removed, and this acted as a stop for the shell in loading. Immediately in front of the shot chamber there was a "grip" in the bore, which was the smallest diameter in its whole length; this was the real calibre of the gun, and the grip was given that the coating of lead on the projectile (as the latter began to move) should be thoroughly compressed into the grooves. From the grip to the muzzle the bore was lapped out, to reduce friction, until at the muzzle it was about ·005 of an inch larger than the actual calibre. The breech end of the A-tube (or barrel) was next bored out and threaded to receive a breech-bush, and the part of the breech-piece in rear of the slot was prepared with a female thread for the breech-screw.

The "vent-piece" (fig. 13) was made of iron or steel (tempered in oil), which, when dropped into the vent-slot or opening in the top of the gun into its position, and pressed by the breech-screw tightly against the end of the powder chamber, effectually closed the bottom of the bore.

All the vent-pieces, except that of the 7-inch gun, were fitted with a copper ring with an angular face, to correspond with that of the copper ring at the end of the powder chamber; the closing of the bore, and, therefore, the safety of the gun, depended upon the exact fitting together of these two pieces. The vent-piece, as its name implies, contained the vent, which descended vertically until it reached the prolongation of the axis of the bore, when it turned at right angles, and lead into the bore. There being no copper facing either to the vent-piece of the 7-inch gun, or copper ring at the bottom of the bore, a tin cup was placed behind the cartridge in order to prevent the escape of gas. The vent-pieces of the 7-inch and 40-pounder guns were fitted with two handles, those of the 20-pounders and smaller guns with one.

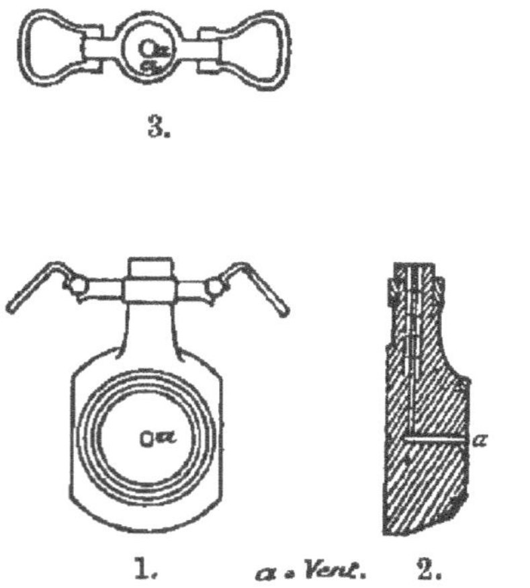

FIG. 13.—VENT-PIECE. 1. Rear Face; 2. Section; 3. Upper Surface.

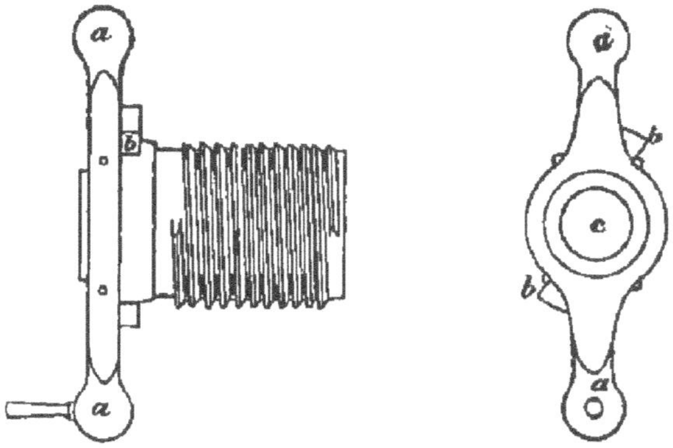

FIG. 14.—40-POUNDER BREECH-SCREW, TAPPET-RING, AND LEVERS.

a. Lever. *b*. Tappet-Ring.
c. Hollow Chamber through Centre of Breech-screw, for Entering Charge and Projectile into Bore.

The "breech-screw" (fig. 14) was made of steel for 20-pounders and smaller guns, of wrought iron or steel for 40-pounders, and of wrought iron faced with steel for 7-inch guns; it fitted into the thread cut in the breech-piece and was worked forwards or backwards by the lever and tappet, so as to press home or release the vent-piece. It was bored hollow, so as to allow of the charge being passed through in loading the gun; the diameter of the hollow being rather larger than that of the powder chamber.

The "tappet-ring" was made of wrought iron, and fitted on to the octagonal part of the breech-screw; it had projections called cams, against which the lever acted, when moving the breech-screw backwards or forwards.

The "lever," also made of wrought iron, fitted on to the circular part of the breech-screw, being kept in its place by two pins working in a groove round the breech-screw.

The first projectiles Lord Armstrong used were of lead, but he afterwards adopted iron coated with lead. The diameter of the projectile including the lead coating, was larger than the calibre of the gun, so that the lead was forced into the grooves, compelling the projectile, when fired, to take the twist of the rifling; thus there was no windage, and the projectile was perfectly centred. After many experiments the projectile eventually adopted had a flat base and ogival head, with a length of body of 3 1 calibres, the radius to which the head was struck being 1·58 diameter, and it does not appear that any improvement on this form has been made since.[1]

The 7-inch gun, which threw a projectile of 110 lb., and was designated the 110-pounder, was first introduced into the Navy in 1861, where it superseded the 68-pounder as the pivot gun of ships, all the Armstrong guns having been proved to possess an accuracy and range far exceeding the best results obtained from the old smooth-bore guns. Not being found, however, sufficiently powerful to penetrate the armour with which all battleships were then being protected, it was replaced in the naval service in 1866 by heavier rifled muzzle-loading guns. The 40-pounder guns were introduced into the service in 1860, and took the place of the 32-pounders as the upper-deck broadside arma-

[1] "Artillery; its Progress and Present Position" (Lloyd and Hadcock).

ment of the larger ships; but a very serious drawback was soon found to exist, which, as a matter of fact obtained in all the Armstrong guns, viz., that there was nothing to prevent the gun being fired before the breech was properly closed, and that unless great care was exercised, the vent-piece was often believed to be screwed close home when this was not the case, in consequence of which several serious accidents occurred. This was especially the case during the bombardment of Kagosima, by the fleet under Sir L. Kuper in 1863, when more than one of the vent-pieces of the *Euryalus's* (the Admiral's flagship) quarter-deck guns were blown out during the heat of the action. In consequence of the unfavourable report sent home by the Admiral, these guns were soon after withdrawn and replaced by what was known as the 64-pounder shunt gun. The smaller natures of the Armstrong guns remained in use in the service for many years after the larger ones were superseded; in fact, the 20-pounder was regarded as the most serviceable gun for use against torpedo boats until the introduction a few years ago of the 3-pounder and 6-pounder quick-firing guns, which have since taken its place. The lighter natures are still in use for land-defences (Great Britain).

Although the original Armstrong guns soon became obsolete, yet Lord Armstrong has had the satisfaction of seeing his method of constructing guns adopted for all the later types which have succeeded his first efforts at producing a satisfactory rifled weapon, and this triumph, even if it stood alone, is a most gratifying tribute to his genius.

Another improvement, which Lord Armstrong effected when his guns were introduced, was in the sights. He made the eye-piece of the tangent scale in the form of a cross slit, with a traversing movement for correcting the effect of a side wind. All the Armstrong guns were sighted, at first, so that the bore pointed to the left of the "line of sight," in order to allow for the constant deviation of the projectile to the right up to a range of about 1,200 yards. Eight minutes deflection was given in the sighting to the 6, 20, 40, and 110-pounder guns, and 12′ to the 12-pounder.

Later the left deflection was not given in the sighting, the sights being adjnsted so that the vertical plane passing through them, when the tangent scale was down, was

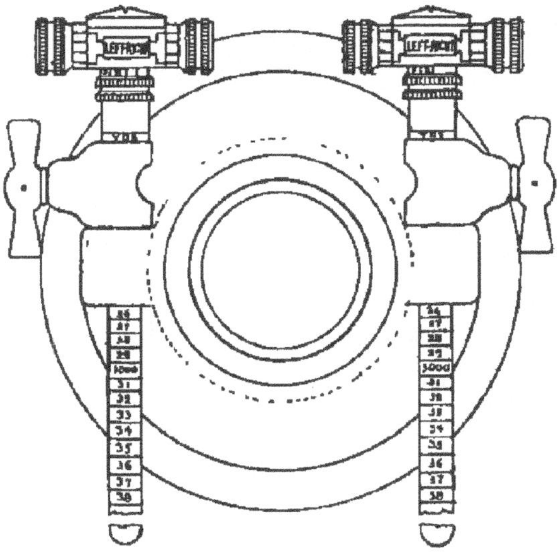

FIG. 15.—ARMSTRONG BARREL-HEADED SIGHTS.

parallel to that through the axis of the bore ; in order, however, to allow for the deviation of the projectile (up to 2,000 yards) the tangent scale was inclined towards the left, at an angle of 2° 16′, and the deflection given in laying therefore increased with the range.

The chief parts of the "barrel-headed sight," as it was called (fig 15), were, the bar, elevating nut, crosshead, two thumb-screws for moving leaf, and the leaf. The bar was made of steel, one side being graduated with degrees, the other with yards. The degrees were divided into six parts of 10′ each; and any number of minutes up to 10′ could be obtained by turning round the graduated elevating nut on the top of the bar, which worked an internal screw and so raised the head of the scale. For instance, if 2° 25′ was the elevation required, 2° 20′ was given by raising the scale, and the additional 5′ by turning round the elevating nut from 0 to 5. The tangent scale or bar, as already explained, was inclined to the left to allow for "deviation," but it was also fitted with a horizontal scale or crosshead, graduated to give $\frac{1}{2}$° of deflection either to the right or left, and this $\frac{1}{2}$° on both sides was subdivided into three parts of 10′ each; at each end of this slide was a graduated nut divided into minutes up to 10′ and these nuts were connected by a screw, crossing the bar at right angles. A leaf with the sight-notch slid along the scale and could be moved right or left by either nut. There was further a sight on each trunnion. The barrel sights fitted each side into what was called the socket-ring, and, when raised, were prevented from sliding down by means of a screw on the side, the socket-ring being secured to the breech by screws.[1]

Shortly after the adoption of his guns Lord Armstrong was appointed Superintendent of the Royal Gun Factories, but he did not hold the appointment long, and he soon afterwards established his factory at Elswick, which has since, under the superintendence of that highly skilled and scientific artillerist Sir A. Noble, developed into those great works only rivalled by Krupp's at Essen.

About the same time that Lord Armstrong was experimenting with his first gun, Sir J. Whitworth also submitted proposals for rifling. He adopted the system of making

[1] "Lectures on Artillery" (C. H. Owen).

the bore of the gun of a hexagonal spiral form, and twisted in the same manner as the Lancaster guns, by which rotation was impressed upon the projectile by rifling "surfaces," instead of by spiral "grooves," and "lands" of a cylindrical bore. The projectiles being of the same hexagonal form externally as the bore was internally, and no forcing process being required, metals of all degress of hardness could be employed, and slowly igniting powder used. Experiments were carried out with three of his guns, 3, 12, and 80-pounders, all breech-loading. The breech was closed by means of a cap, which worked in an iron hoop, jointed to a projection at the side of the breech, and which, when turned to its proper place, was screwed externally to the breech-piece, by three turns of the handles, which worked the screw. To open the breech, the cap was unscrewed and turned aside, leaving the bore of the gun open and clear from end to end.

The guns were all made in masses of homogeneous iron and bored out, the large guns being strengthened by wrought-iron hoops forced on by means of hydraulic pressure. The projectiles were uncoated, hard metal bolts of various shapes, according to the purpose for which they were to be employed, and were so nicely shaped that their bearing surfaces fitted with the utmost exactitude. For piercing armour plates, the projectiles were made of homogeneous iron, and were flat headed; where length of range was important, the fore part of the projectile was made to taper slightly, the head being rounded off, and the rear parts made nearly to correspond with the fore in regard to the degree of taper, but the base was flattened and sometimes slightly hollowed out. With his 3-pounder, and an elevation of 35 degrees, a range of 9,463 yards was obtained; with the 12-pounder, and an elevation of 10 degrees, a range of 4,059 yards was obtained, while with the 80-pounder and a similar elevation, the range was 4,409 yards; when fired against the floating battery, *Trusty* which was protected with $4\frac{1}{2}$-inch wrought-iron plates, in each of the four rounds the 80-pound projectile passed completely through, making a clean hexagonal hole.

"In 1863[1] a competitive trial was ordered between Sir J. Whitworth's and the Armstrong guns. The Whitworth guns were now made, however, of mild steel, loaded from the muzzle but still rifled on the hexagonal principle. The 12-pounder submitted was forged in one piece, with a hoop over the powder chamber. The 70-pounder had an inner tube, which was screwed into the body, and was further strengthened with hoops, forced on by hydraulic pressure when cold. Lord Armstrong submitted muzzle-loading guns, as well as his previously described breech-loaders, all having steel barrels, and all built up on the Armstrong principle." The rifling of the muzzle-loaders was on the shunt system, which is explained on p. 51.

Referring to the strength of the guns, the Committee reported as follows: "After, then, these 12-pounder (8 cwt.) guns, viz., the Armstrong muzzle-loading and breech-loading guns, and the Whitworth muzzle-loading guns had each fired 2,800 rounds, the M. L., with 1 lb. 12 oz. of powder and 12 lb. shot, attempts were made to destroy them by firing increased charges of powder and shot, with the following result: at the forty-second round, the Armstrong breech-loading gun split open, but did not burst; at the ninety-second round the Whitworth steel gun 'burst violently' into eleven pieces, while the muzzle-loading Armstrong gun failed at the sixtieth round, one of the outer coils having cracked and fallen off without flying to pieces." The result of these trials was the adoption of

[1] Lloyd and Hadcock, "Artillery, its Progress and Present Position."

the Armstrong system of building guns up, although certain modifications were adopted by Woolwich later on, to which we shall refer in due time.

By 1864 the use of armour-plating for vessels of war had brought the question of penetration and power of guns prominently forward, and the Admiralty proposed the construction of large smooth-bore "wrought-iron" guns. Two natures were accordingly made, 100 and 150-pounders, which were built up on the Armstrong system. More powerful results, however, were obtained from the smaller rifled guns, with which experiments were being carried on at the same time, and it became clear that no smooth-bore weapon could hope to compete with rifled guns, so the manufacture of smooth-bore guns was soon after discontinued, the 100-pounders being the last muzzle-loading ordnance of this nature introduced into the service, and for all practical purposes they soon became obsolete, and the rifled gun became fairly established as the weapon of the future.

CHAPTER III.

THE ADOPTION OF MUZZLE-LOADING RIFLED GUNS, THEIR CONSTRUCTION, RIFLING, MOUNTING, ETC.

AFTER having been in use for over five hundred years, smooth-bore ordnance were in 1864-65 finally condemned, to make way for rifled guns. The general introduction, some twenty years previously, as mentioned in Chapter I., of shell-guns, the destruction of the Turkish Squadron at Sinope on November 30th, 1853, by the shells from the Russian ships, and the great damage inflicted on the vessels of the allied fleets at the bombardment of Sebastopol by the shells from the forts, were the causes which led to the introduction of armour, as the only method of defence which would protect ships from these terrible missiles.

It very soon became evident that the heaviest spherical projectile from a smooth-bore gun was quite inadequate to pierce the newly-adopted armour plates, and artillerists speedily began to devote their attention to producing weapons which should be powerful enough to again turn the scale in favour of the gun and against the ship. With the definite adoption of the muzzle-loading rifled gun in 1864 began the great contest between guns and armour which has lasted ever since, and is being carried on with as much energy as ever at the present day. The only form of projectile which could be used from smooth-bore guns was the spherical, a shape, which it is easy to understand did not lend itself to the penetration of even the, comparatively speaking, thin iron plates which were first introduced; while the range of the heaviest of these guns was limited and the shooting inaccurate. In all smoothbore guns the windage, that is, the difference between the diameter of the bore and that of the shot was considerable, and the result was: (1) the loss of a certain portion of the force of the charge, due to the escape of the gas over the projectile; (2) irregularity in the flight of the projectile. With regard to the first point, the loss of a certain portion of the gas meant a loss of propelling force, and a consequent proportionate loss of initial velocity. The irregularity in the flight of the shot arose from the fact that the centre of the ball was below the axis of the piece, and therefore the gas acted first upon its upper portion, driving it against the bottom of the bore; as it was impelled forward by the charge, the reaction drove it against the upper surface of the bore further down, and so on by a series of rebounds, until it left the gun, when its flight was determined by the chance as to where it last struck, for if this happened to be on the right side of the bore, it would glance off that side, and leave the gun with what was called left deflection, or *vice versâ.* The shortness of the range was due, partly to the bad form of the spherical shot for overcoming the resistance of the air, and partly to the want of weight entailed by that form, aud experiments soon showed that only to projectiles of an elongated form, combining as much weight as possible with small diameter could the velocity and penetrating force be given, which could cope successfully with the new defensive powers afforded to ships by armour-plating. In order, however, to obtain the range and precision of fire, it became necessary to devise some means for keeping the elongated projectile point foremost in the air, as they have a tendency to turn end over end, as the pressure of the air acts unequally upon them when they have left the bore of the gun, owing to the centre of gravity not being in the same spot as the centre of sectional area. To counteract this tendency, it has been found that by elongating the point of the projectile, and giving it a spin or rotation about its longer axis, it travels through the air with its point always approximately first, and to attain this end the projectile is given a spin or rotation by the rifling in the gun, while elongating the head gives a less cross-sectional area to the resistance of the air. It will thus be seen how rifled guns became a necessity.

In the last chapter the method adopted by Lord Armstrong in his first rifled guns was fully described, and also that a number of accidents, arising from the vent-pieces not being properly screwed up, a defect to meet which no satisfactory remedy was apparently devised, led the authorities to discontinue the issue of the larger nature of these guns. After a long series of experiments muzzle-loading guns, rifled on the so-called Woolwich system, were adopted in 1865; the rotatory motion being at first and for many years imparted to the projectile by means of rows of studs on its body, and by cutting spiral grooves, termed "rifling," in the bore of the gun into which the studs on the projectile fitted. The great disadvantages attaching to the stud system were, that in the first place, there was a considerable amount of windage, as it was necessary that the projectile should be able to pass freely down the bore of the gun when loading; and secondly, that the bodies of projectiles and of shells in particular, were unavoidably weakened by the insertion of the studs. In the muzzle-loading rifled ordnance there are six distinct forms of groove used for rifling, viz., the Shunt; Plain; Woolwich; French; French modified; and the M.M., or Maitland muzzle-loading groove.

When rifled guns came into use there existed in the armament of all nations a number of smooth-bore cast-iron guns, and, for the sake of economy, it was attempted to turn them into rifled pieces. The material, however, was found too weak of itself, and different modes of strengthening the guns were tried.

As early as 1855 attempts were made to strengthen such guns by encasing them in rear of the trunnions with a wrought-iron or other jacket, and between that date and 1863 various plans of this and other descriptions were tried unsuccessfully. In that year the late Sir William Palliser proposed to line cast-iron guns with coiled iron barrels, fitting comparatively loosely into the casing until expanded by the heavy proof rounds. This method appeared to be more promising than any previously tried, and was moreover founded upon correct principles, the stronger material being placed next the charge.

Several guns converted on the Palliser principle gave very fair results upon trial, and showed themselves more powerful than the smooth-bores from which they were made. It was consequently determined to convert a large number of smooth-bore ordnance in this way, for although such guns would be much inferior to built-up Woolwich ordnance, yet where the range was limited, and there were no iron plates to pierce, they would prove useful.

The Ordnance Select Committee proposed several natures for conversion, but it was finally decided that only the following natures of smooth-bore guns should be converted :

68-pr. of 95 cwt., S.B.,	into	80-pr.	R.M.L., of	5 tons.
8-inch of 65 ,, ,, (throwing 50 lb. shell)	,,	64-pr.	,,	71 cwt.
32-pr. of 58 cwt., S.B.	,,	64-pr.	,,	58 ,,

The mode of conversion consisted in boring out the old

gun and making a wrought-iron tube to fit the casing thus prepared; this tube was slightly smaller than the bore of the casing, and was pushed into it without much force being necessary. After proof, however, the barrel being permanently expanded fitted tightly against the interior of the casting. When fitted into its place it was secured there by means of a cast-iron collar screwed into the muzzle end of the casing over a shoulder on the end of the tube; a wrought-iron plug was also screwed through the casing underneath and into the barrel, preventing any chance of the latter shifting round.

The "Shunt" groove was adopted on the first introduction of rifled muzzle-loading guns, and the object of the "Shunt" was to centre the projectile (that is, to make its axis coincide with that of the gun) before it left the muzzle. The groove, which had a uniform twist of one turn in forty calibres, varied in depth and width (fig. 16), and was wider than the studs on the projectile, one side being deeper than the other, so that as the projectile was rammed home, the studs moved freely on the deeper side, but when the gun was fired, the projectile rotated in the opposite direction, the studs shunted over up an inclined plane which led to the shallower part of the groove; by this arrangement the projectile was slightly gripped on three sides at once, and centred in the bore before leaving the muzzle. This system, however, was not a success, as it was too compilcated, the strain was often too severe, the studs gave way, and the projectile left the gun with imperfect rotation, which materially affected the accuracy. The only guns rifled on this system were the 64-pounders, which were the first muzzle-loading rifled guns constructed, their adoption being decided upon in 1864; but

FIG. 16.—THE SHUNT GROOVE.

even with them the "Shunt" system was soon given up for the "Plain" groove, which was practically the deep portion of the shunt; the bottom of the groove being concentric with the bore, and the sides formed by straight lines at the same outward inclination from the bottom of the groove (fig. 17). These guns are now quite obsolete, but at one time a large number were issued to the Navy, and they formed the principal armaments of the wooden frigates and corvettes of twenty and twenty-five years ago; they were built up on the Armstrong pattern, with a barrel or inner tube of coiled iron, a forged breech-piece, a B-tube with a muzzle-piece for the swell, four external thin coils, a trunnion ring of forged iron, and a cascable screwed into breech to support a copper cup, which closed the end of the bore (fig. 18). In 1871 solid-ended steel barrels were adopted instead of the coiled iron tubes, and a jacket with a triple coil over the breech was substituted

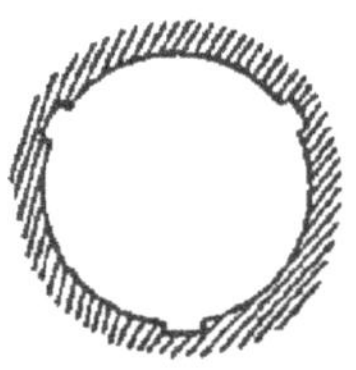

FIG. 17.—THE PLAIN GROOVE.

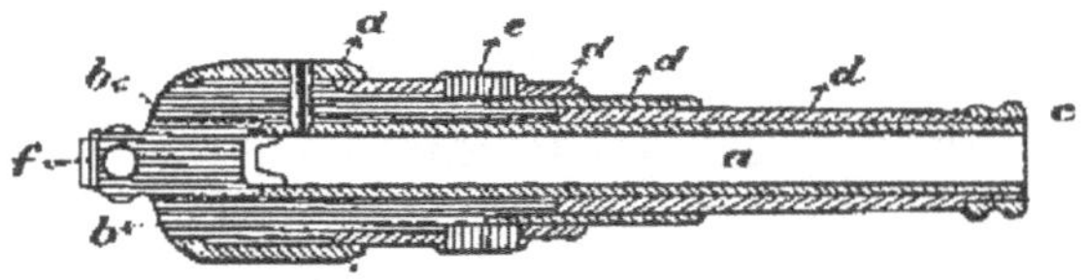

FIG. 18.—64-POUNDER GUN. EARLIEST PATTERN.

a. Barrel. *b.* Breech-piece. *c.* B-tube.
d, d, d, d. External Coils. *e.* Trunnion Ring. *f.* Cascable.

for the forged breech-piece, a B-tube of coiled iron being shrunk over the chase (fig. 19). This alteration in the construction much increased the strength of the gun, so that it was possible to use both a larger powder-charge and projectile, the weight of the latter being increased from 64 lb. to 90 lb., although the gun was still known as the 64-pounder.

In 1865 the first heavy rifled guns, which marked the new departure in gun-making, were constructed; they weighed 6½ tons, had a calibre of 7 inches, were 10 feet 5·25 inches long, and threw a solid projectile of 115 lb. in weight. These guns were followed almost immediately by 9-inch guns, weighing 12 tons, with a 256 lb. projectile.

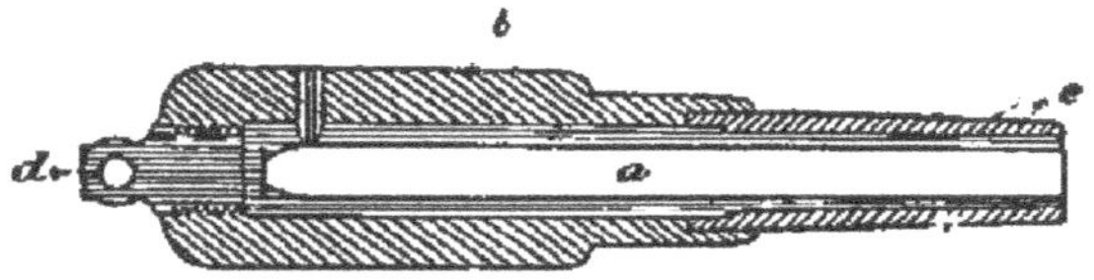

FIG. 19.—64-POUNDER GUN. LATEST PATTERN.
a. Barrel. *b.* Jacket. *c.* Outer Coil. *d.* Cascable.

In 1868 a further advance was made in the production of the 10-inch 18-ton gun, throwing a projectile of 410 lb.; and in 1870 came 12-inch 25-ton guns with a projectile of 614 lb. Next year a 12-inch gun of 35 tons was produced, and in 1874 a 12·5-inch gun of 38 tons, both these last throwing a projectile of 820 lb. in weight. All these guns were rifled with the "Woolwich" groove, which differs from the plain in being rounded off on both sides to avoid sharp angles, and prevent any tendency in steel to split along the edge of the groove (fig. 20), but the number of the grooves, and the amount of twist in the rifling, differed according to the nature of the gun. In the 7-inch gun the rifling consists of three grooves, 1·5 inch wide and ·18 inch deep, with a uniform twist of one turn in 35 calibres. It should be stated that there are three systems of twist in rifling in common use: the "uniform," the "increasing," and a combination of the two. Uniform twist possesses the advantage of simplicity, the bearing parts on the projectile may be distributed over the whole body of the

FIG. 20.—THE WOOLWICH GROOVE.

shot, and the projectile leaves the muzzle with steadiness of flight. Experiment has shown that, other conditions being similar, a gun with uniform twist will give greater velocity to its projectiles than one with increasing twist; the shot having to move forward, and, at the same time, take the whole turn of the rifling. It does so at first, comparatively speaking, slowly, and the powder is therefore given more time to burn, and consequently develops more energy; but this puts a severe strain on the gun, and there is a tendency, moreover, to shear the soft metal of the studs, or else for them to over-ride in the grooves, and, to remove this disadvantage and risk, it was determined to adopt a groove with an increasing twist, which has the advantage of putting less strain on the gun, as the full rotation is given to the projectile gradually, there being an imperceptible change from little or no twist at all at the commencement of the rifling, up to the maximum twist at the muzzle. It must be said, however, that this brings a heavier strain on the muzzle portion of the gun, where there is least strength to resist it. In all modern guns the advantages derivable from each nature of twist are secured by a combination of the two; the twist may commence at the bottom of the bore from zero, or with a definite pitch (now always preferred), and at some point in the bore, when the inertia of the projectile has been overcome and the increase of velocity is small, the curve is exchanged for a straight line representing uniform twist. With muzzle-loading guns, this portion of uniform twist was necessarily short, and the combination will not be found where the rifling was made for studded projectiles. In the 9-inch 12-ton guns, the rifling consists of six grooves, 1·5 inch wide and ·18 inch deep, with a twist increasing from 0 to one turn in 45 calibres at the muzzle. In the 10-inch 18-ton guns there are seven grooves, 1·5 inch wide and ·2 inch deep, the twist increasing from one turn in 100 to one in 40 calibres, the maximum being reached at the muzzle; the 12-inch 25-ton gun has nine grooves, 1·5 inch wide and ·2 inch deep, the twist increasing from one turn in 100 calibres to one turn in 50 at the muzzle. The 12·5-inch 35 and 38-ton guns have nine grooves, 1·5 inch wide and ·2 inch deep, with a twist increasing from 0 at the bottom of the bore to one turn in 35 calibres at the muzzle.

In 1873 it was proposed to construct a still more powerful gun, and in 1875 the first 80-ton gun was completed, with a calibre of 14·5 inches. After a series of experiments it was bored out to 15 inches, and after further trials was again bored out to 16 inches, which became its final calibre. It was at first rifled on the Woolwich system, but it was afterwards determined to adopt the polygroove, the Maitland muzzle-loading system of rifling being chosen, which consists of a large number of very shallow grooves (fig. 21). The bottom of each groove is concentric with the bore, and the sides terminate in quadrants of a circle. There are thirty-three grooves, 1 inch wide and ·1 inch deep, while the twist increases from 0 to one turn in 50 calibres. The principal advantage of polygroove rifling is, that the gun is not so much weakened by a number of shallow grooves as by a few of the necessary depth to accommodate studs of sufficient strength to impart rotation to the projectile. The projectiles for this gun are not studded, but the rotation is given by means of what is called a gas-check, the object and use of which must now be explained.

FIG. 21.—MAITLAND MUZZLE-LOADING GROOVE

The adoption of muzzle-loading rifled guns with studded projectiles was accompanied by one great evil, viz., the windage over the body of the projectile. It has already been explained that windage is the difference between the diameters of the projectile and the bore of the gun, it being necessary to leave a space in order to facilitate loading. The gas caused by the explosion of the charge rushing through this space cut its way into the steel barrel, and caused what is called erosion, which naturally became worse the more the gun was fired. The employment of large charges of powders possessing great specific heat increased the evil, which also affected the range of the guns, as even a small amount of windage wastes a certain proportion of the energy developed by the powder. It was first attempted to get rid of this erosive action by means of wads, made of *papier-maché*, and then by means of what were called "Bolton wads," but both proved quite useless for the purpose, and, after many experiments, a gas-check made of copper, to which three per cent. of zinc had been added, was finally introduced. The gas-check, the idea of which emanated from the Elswick firm, is a copper disc, extending completely over the base of the projectile, and somewhat resembling a soup plate, and is attached to the base of the shot before loading, with which operation it does not interfere, but when the gun is fired the pressure in rear expands it against the bore and thus checks the escape of the gas. The earlier gas-checks were smooth, as the rapid action of the powder then in use, known as the R.L.G., or rifle large grain, expanded them into the grooves of the gun almost instantaneously; but when slower-burning powders were introduced, the expansion of the gas-check was more gradual, and a certain quantity of gas escaped, so that some scoring still continued. To obviate this, gas-checks with ribs or projections to fit the grooves of the guns were introduced; for guns having the Woolwich rifling these ribs are cast and planed; for those rifled on the polygrooved system, the spaces for the lands are simply planed out of the continuous edge.

The 80-ton gun is the largest muzzle-loading gun for the Navy (fig. 22) which has been constructed; there are four of them in use, and these are mounted in the turrets of the *Inflexible.* The gun fires a projectile 1,700 lb. in weight, and has an extreme range of about seven miles. In 1878 four 100-ton rifled muzzle-loading guns with a calibre of 17·5 inches, built at Elswick, were bought by the Government, two are mounted at Malta, and two at Gibraltar. The *Duilio* and the *Dandolo,* two of the earlier large Italian

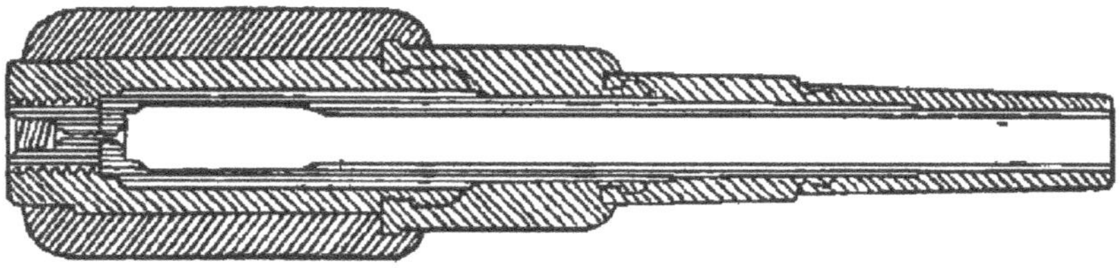

FIG. 22.—SECTION OF 80-TON GUN, SHOWING METHOD OF BUILDING UP.

battleships, each carried, when they were first built, four of these guns, two guns being mounted in each turret.

Although rifled muzzle-loading guns have become obsolete, yet they still form the heavy armament of many of our older battleships. In giving a description of the manufacture of this class of ordnance, it will only be necessary to take the construction of one gun as a type, and what is known as the 7-inch Mark IV. is chosen for the purposes of illustration in the official treatise on ordnance as being one of the latest designs among rifled muzzle-loading guns.

The A-tube, or barrel, was made of steel, which came to the Royal Gun Factory in the form of a solid forged steel ingot. Small pieces were first cut off for purposes of testing, which were supposed to give from 34 to 46 tons breaking strength, 21 tons limit of elasticity, and 17 per cent. elongation in two inches. If the block passed all the tests, it was then bored out and tempered, or hardened, as the process ought more properly to be called. For boring, the ingot was placed in a lathe, in which the tool is practically stationary, while the ingot was made to revolve against the cutter; in this way the momentum of the mass was made to assist in the work of cutting out the interior metal. It was next annealed, by being heated and allowed to cool very slowly; the object being to remove any internal strains produced by forging.

Tempering, or hardening, was then carried out; the barrel being placed in a vertical furnace (the bottom resting on supports), only wood being used for fuel, as it is cleaner, and when heated to a proper temperature (which was judged by experience) it was lifted out by an overhead crane and lowered into a tank of rape oil, where it was allowed to cool slowly for some seven or eight hours, water being kept circulating round the tank in order to cool the oil. Hardening causes internal strains, for the surfaces of the metal cool more rapidly than the interior; the latter, therefore, as it contracts is put in tension. As these strains are dangerous, the barrel was again annealed, which was done by heating it again in the same furnace to about 900 degrees Fahrenheit, a dull red heat, judged practically as the temperature at which a piece of deal wood held against the metal will just be charred, the barrel being then allowed to cool very slowly; after this it was again slightly turned and bored out to remove surface cracks, and then subjected to a water test. For this it was placed in a hydraulic press, the muzzle end being tightly closed and made water-tight, and water was then forced in from a hydraulic main until a pressure of four tons on the square inch was registered, when, if no flaw could be detected by moisture on the exterior, the barrel was considered sound.

It has already been stated that, as the result of a long series of experiments, the principle of building up guns, introduced by Sir W. Armstrong, was finally adopted for the new ordnance. Various modifications and improvements however were made, and the system finally adopted was one suggested by the late Mr. W. Fraser, Deputy Assistant Superintendent of the Royal Gun Factory, who advocated the use of a few large hoops (instead of several small ones) by forming two thick coils, one upon the other, without separate welding or machining, since termed the "Fraser" construction. The 7-inch gun built on this principle consisted of a barrel, the method of manufacture of which has just been described, a breech-piece, B-coil, B-tube, jacket, and cascable.

The breech-piece was made of coiled iron; a long iron bar being used for the purpose, which, after being brought to the necessary heat, was drawn out and the end hooked to a pin on a mandril, which was then revolved, drawing the bar out, and at the same time coiling it round itself. The bars ranged in length generally from 150 to 200 feet. After coiling, the next operation was to weld the coil to unite all the folds and make a solid cylinder of iron. For this it had to be replaced in a furnace and brought up to the necessary heat, it was then put under a steam hammer and a few blows were sufficient for welding, it was then reheated and the operation repeated at the opposite end. Double coils were made by winding a second bar round the cylinder formed by the first, but this process was only carried out in the case of very heavy guns (fig. 23).

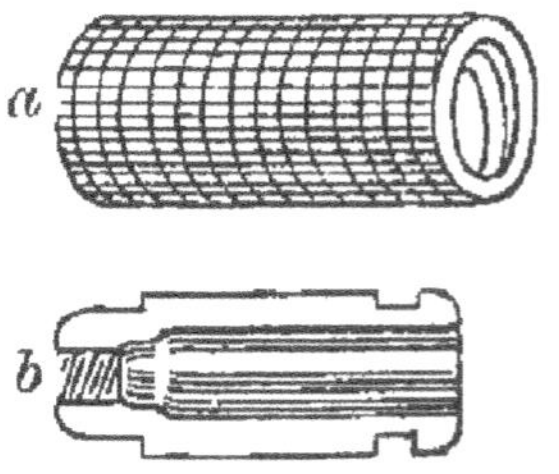

FIG. 23.

a. Breech-piece in Coil.
b. Breech-piece ready for Shrinking.

The trunnion ring was formed from a solid forging, two

holes being punched in the centre, which were gradually made into one, until a fairly shaped ring was completed; in this way the fibre of the iron, which in forging takes the direction of the length, is distributed round the ring without severance, and along each arm of the trunnions (fig. 24).

The jacket, or outside breech portion, was made by welding together two coils and the trunnion ring at one single heat, the three parts being rough shrunk together, the surfaces in contact being previously turned in a lathe. The diameter of the trunnion ring when cold was smaller than the diameter of either part over which it would rest, so that, after cooling and shrinking, it was known there

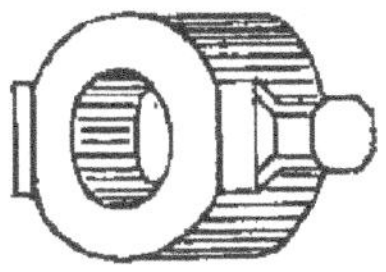

FIG. 24.—TRUNNION RING.

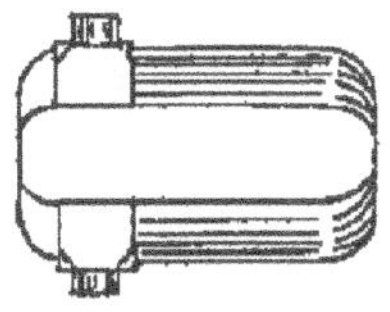

FIG. 25.—TRUNNION RING AND JACKET.

was sufficient metal at the joint to insure perfect contact and welding. The trunnion ring in this way formed a band over the joint, and the operation of welding the jacket was conducted in much the same manner as that of welding a coil (fig. 25).

In building up guns, the various parts have always been shrunk on, as it is an easy and effective method of binding the parts firmly together, and all the metal is made to contribute to the strength of the gun in fair proportion; but in 1866 an important improvement was made by the introduction of what was called the "Hook Joint" for rifled muzzle-loading guns from the 64-pounder upwards. The front of the breech-coil inside had left on it a ring of slightly smaller diameter than the coil, which ring fitted into a recess turned on the tube or coil below it; and although able to pass over the coil when heated, on cooling the excrescence locked into this recess, and so formed an additional grip to hold the parts together. The operation of shrinking is simple; the outer coil is expanded by heat until sufficiently large to fit easily over the inner. When raised to a proper heat the coil is lifted from the fire, and the inside being thoroughly cleaned, it is dropped over the barrel or gun (set vertically in a pit) which has been prepared to receive it. The heat required is not very great, for the linear expansion of iron between 32 degrees and 212 degrees Fahrenheit has been determined by several authorities at 0·00122 of its length, so the co-efficient of expansion for every degree is abont 0·000007. It is calculated, roughly speaking, that the diameter of a coil of wrought iron would be increased to the extent of one-thonsandth by the addition of 150 degrees of heat. Some 600 or 800 degrees are therefore sufficient for the operation of shrinking. Great care has to be taken in cooling, and the mass must be allowed to cool slowly and shrink towards the end at which contraction is allowed to commence, water being generally applied to the part where first contact is wanted, which is the shoulder of bearing; the inside of the gun is always kept cool, to hasten the work and promote regular contraction, commencing from the inside of the coil, rings of gas jets being used outside to keep the heat upon the outside while the inside cools.

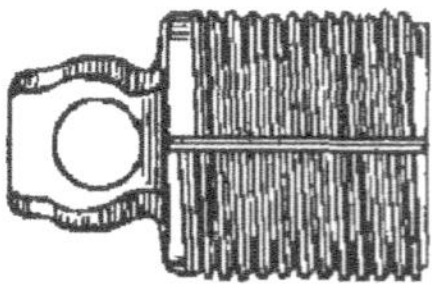

FIG. 26.—THE CASCABLE.

The cascable (fig. 26) was the last piece of the gun to be fitted, and it was screwed into the breech after the gun had been otherwise built up. It was forged solid, turned into shape and fitted with a strong bevel thread, the end being shaped into a loop, through which the breeching was rove. Care had to be taken that the front bore evenly against the end of the barrel, and before being finally screwed in a channel was cut longitudinally on the right side, forming what is called the gas-escape, which is intended to act as a tell-tale in case the steel barrel should split. After being screwed into its place, a plug was inserted through the breech-piece into the cascable to prevent its getting loose. The bore was then finished to calibre by fine boring, finished boring, broaching, and lapping.

The gun was now rifled, and for this purpose it was placed horizontally in the machine, when each groove was cut singly, the gun being turned on its axis after each groove was finished, to allow of the next being cut. The rifling bar is turned in the axis of the gun by means of rack-work and a copying-bar cut to the curve that is required for the particular form of twist. If the twist is uniform, a straight inclined bar is all that is required.

The guns are vented so that the cartridge should be ignited at about four-tenths of its length from the end of the bore, as from experiments it was ascertained that the best ballistic results were obtained when the cartridge was ignited almost at the middle. The vent is radial in most rifled muzzle-loading guns, but axial, *i.e.*, in prolongation of the axis, in the 12·5-inch gun and upwards, and is placed in the middle of a copper bush, the gun being bored out to receive it; this bush is a cylinder of hardened copper $1\frac{1}{8}$ inch in diameter, which is made to screw into the gun, its length varying according to the thickness of metal of the gun.

The gun had next to be proved, and if it passed successfully through the tests it was finished off ready for issue. The most important of the final operations for completion was the fixing of what is called the line of metal, which is marked on the upper surface of the breech and muzzle by lines about two inches long, this line being the basis of all sighting arrangements. The sighting has to be carried out with great accuracy and care; holes have to be drilled for the sight-sockets, and allowance has to be made, when doing this, for the permanent correction for drift, which has to be given to the hind sights; as the rotation of the projectile gives a deviation to the "right" in British guns, owing to the rifling being "right-handed," but as this is practically constant, it is allowed for by inclining the rear or tangent sights. A heel scale is engraved on the cascable for all guns used in the naval service, for which wood scales are provided, the scale reading to about three and a half degrees above and below zero. The broad arrow and actual weight are stamped on the breech, also the centre of gravity and half weight mark on the heavier guns. Other marks are end of bore, end of rifling; and on the left trunnion is found (1) the name of the firm or factory where the gun was made; (2) the registered number of the gun; (3) the numeral indicating the pattern or mark; and (4) the year in which made. Thus

means, manufactured at Royal Gun Factory, No. 301 Mark III., pattern in 1868.

The above-described method of constructing guns proved eminently successful. The inner steel tubes gave the strength, hardness, and freedom from flaws which was so important for rifled barrels; steel, however, in its early days of use for guns was difficult to work, and in France and Germany, where from the first introduction of rifled guns steel was used entirely for their manufacture, several serious accidents, attended by loss of life, occurred at various times through guns bursting, as the steel fragments flew in all directions. The coiled iron outer tubes and jacket, used in England, proved a great security against accidents of this kind, as although less hard, the iron was more elastic than the steel, and even where, as happened occasionally, the inner steel tube split, the outer coverings remained sound and no harm was done, there being only one instance of a Fraser gun on service bursting into pieces and causing loss of life, when one of the 38-ton guns mounted in the foremost turret of H.M.S. *Thunderer* burst at target practice, killing two officers and eight men, besides wounding several others. In 1869, in some experiments which were carried out with a 9-inch gun, the steel tube was split at the 1,008th round, but it yielded gradually, and the strength of the jacket was thoroughly proved by firing forty-one rounds after the barrel had split, and still the exterior remained perfectly sound. The accident on board the *Thunderer,* which occurred on January 2nd, 1879, was clearly shown before the commission of officers and scientific experts appointed by the Admiralty to investigate its causes, to have been due to no fault in the gun but to have been the result of a deplorable error in loading. It was proved beyond doubt that the gun had been fired when loaded with two complete rounds of powder and projectile. The gun formed one of a pair in the foremost turret, and was run in and out by hydraulic machinery. During the practice, both guns being loaded with a battering charge of 110 lb. of powder and a Palliser projectile 800 lb. in weight, it was intended to fire the two guns simultaneously. One gun, however, missed fire, although nobody apparently was aware of the fact; the concussion and the smoke made by the discharge of one gun are so great that it would be hard, unless anyone was on the look-out, to tell that they were not due to both guns. The men

Instantaneous Photograph; taken for Sir W. Armstrong & Co., Elswick. *To face p. 64.*

H.M.S. "VICTORIA" FIRING ONE OF HER 111-TON GUNS, AT HER TRIALS OFF THE TYNE.

were drilled to run their guns in directly they fired, so as to be in readiness for loading, and not observing that one of the guns had not gone off, they proceeded to run in, and both guns were reloaded, with the result that the one with the double charge burst when fired next round.

When the Admiralty received the report of the commission they determined, in order that no question should arise in the future as to the correctness of the conclusions which the committee had arrived at, to have the other gun brought home and subjected to various trials, with a view of ascertaining under what other conditions it would burst; but after a series of experiments, when every suggestion had been tried without injuring the gun, it was double loaded, just as the commission reported that in their opinion the burst gun had been. When the smoke cleared away from the cell in which the gun had been placed for these trials, it was found that it had burst like its former companion, and that the fractures in the two guns were nearly identical.

This accident undoubtedly strengthened the feeling in favour of a return to breech-loaders which had been growing in the mind of the Navy for some time, since it was quite certain that no such mistake as double-loading could be made with a breech-loading gun, as it was impossible to force the projectile home beyond a certain distance, and consequently there would not be room for a second charge.

Another accident, which, however, fortunately was not attended by loss of life, occurred to one of the Elswick 100-ton rifled muzzle-loading guns mounted on board the Italian battleship *Duilio* in March, 1880, when the gun was being tried for acceptance. In this case the barrel was fractured, the rear portion being forced to the rear, and carrying with it the covering jackets, none of the superimposed coils, however, being injured.

Commander Lloyd and Mr. Hadcock, in their work on "Artillery," ascribe the accident to the following causes: "It was principally due to the peculiar condition of the powder charge, which was composed of 550 lb. of Fossam powder, a powder that may be described as having a slow-burning exterior and a quick-burning interior—a combination which, in theory, should, with exterior ignition to the cubes or grains, give very high ballistic results.

"It is believed that the powder cubes had on the occasion of the accident got broken, and the charge was thereby converted into a very violent one, capable of giving enormous pressures, and as the gun was undergoing proof at the time the charge was heavier than would have been used under service conditions.

"The form of the front end of the chamber may also have been conducive to the accident, for it altered too suddenly. This is shown by the fact that the rupture was an exceedingly clean one, and took place exactly along the line of alteration to form. It is worthy of remark that this gun did not burst explosively as was the case with the *Thunderer's* gun, and it was repaired by its several parts being built up again over a new barrel."

It is time now to turn to the question of the mountings, which, with the advent of heavier guns, necessarily took the place of the truck carriages already described.

For pivot-guns a wooden carriage mounted on a wooden slide, which traversed on gun-metal racers secured to the deck, had long been in use, and with the advent of first the 68-pounder, and then the 7-inch gun for broadside purposes in the earlier ironclads, it became necessary to mount them also on carriages and slides. A very brief description of the earlier form of carriage and slide is necessary. The carriage itself was almost similar to the truck-carriage, but instead of trucks had flat wooden blocks, faced with iron,

Fig. 27.—100-Pounder Smooth-Bore Gun, Mounted on Wooden Carriage and Slide.

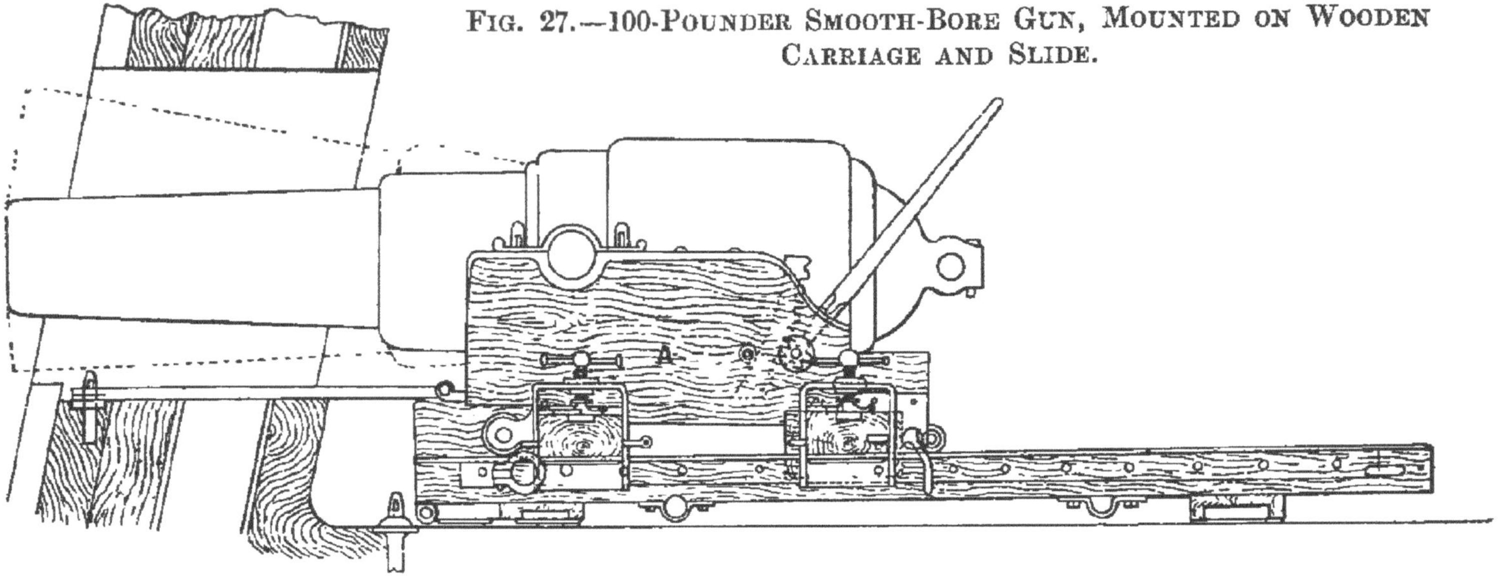

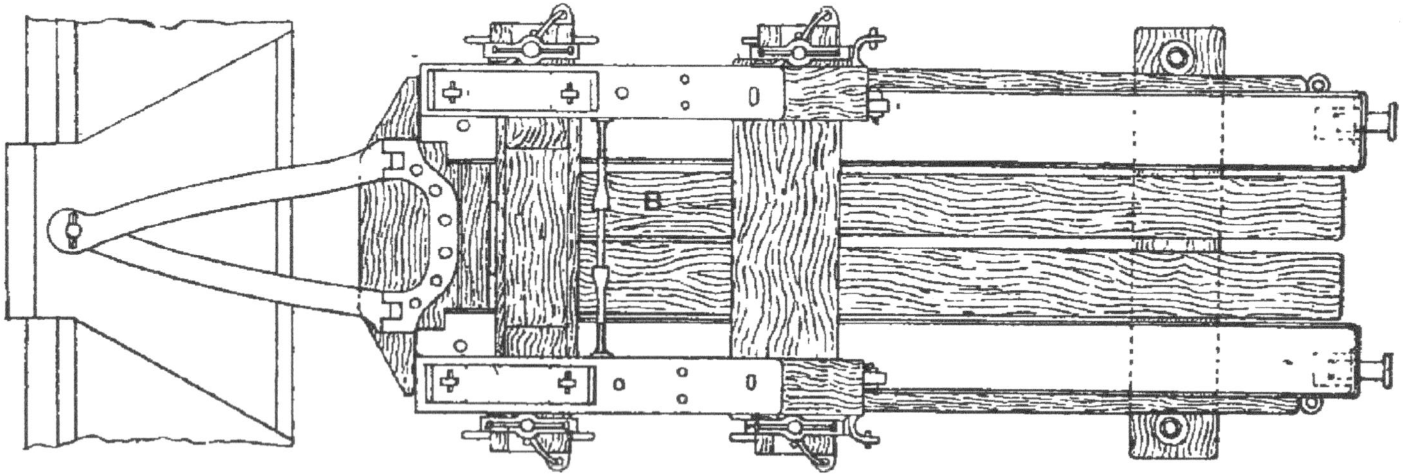

Fig. 28.—Wooden Slide for Heavy Guns.   B. Carriage.

both in front and rear, which rested on the slide; there were two small gun-metal trucks or rollers attached to the front of the brackets, and two smaller trucks still in rear, which worked on eccentrics, so that the carriage could be thrown on and off them as requisite for running in or out by means of levers working in sockets attached to the eccentrics; when the gun was fired, it recoiled on the blocks, the rollers not being brought into play (fig. 27).

The slide (fig. 28) consisted of two side pieces connected by a head-block and two under chocks (blocks of wood shod with metal friction-plates, to take the bearing upon the racers); iron plates were secured on the top of the side pieces on which the rollers (or trucks) and the blocks of the carriage worked when the gun was running in or out. The slides could be pivoted both in front and rear, and, as already stated, the under chocks traversed on racers let into the deck of the ship. The carriages were fitted with an iron compressor (fig. 29) on each bracket of the carriage, which, when screwed up, checked the recoil of the carriage on the slide. The guns were first fitted to elevate by means of an elevating screw under the breech, then an iron circular band was placed round the gun between the brackets, which raised or lowered the gun by means of a pinion fitting into a toothed arc attached to the inside of the bracket and worked by a lever outside the bracket. When there was no recoil, the gun was run in by what was called the preventer rope, after being first thrown on the rollers by the eccentrics; it was run out by side ropes, the preventer rope being used as a check when necessary. The

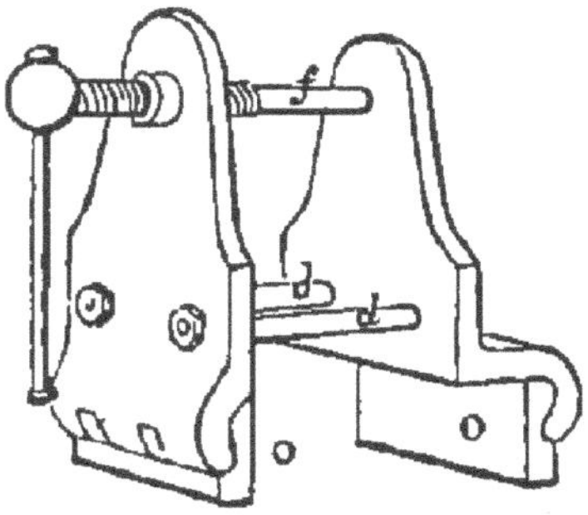

FIG. 29.—COMPRESSOR.

slide was trained by means of tackles hooked to the ship's side, handspikes being also used. This method of mounting and training guns was a great improvement on the old truck-carriages, but a further advance was soon made, and iron carriages and slides took the place of the wooden ones; and in 1864 what is known as the R. C. D.[1] carriage and slide, which are still used for 7-inch and 9-inch muzzle-loading guns, were finally adopted (figs. 30 and 31). The most important of the many improvements embodied in the new carriage and slide again emanated from the Arm-

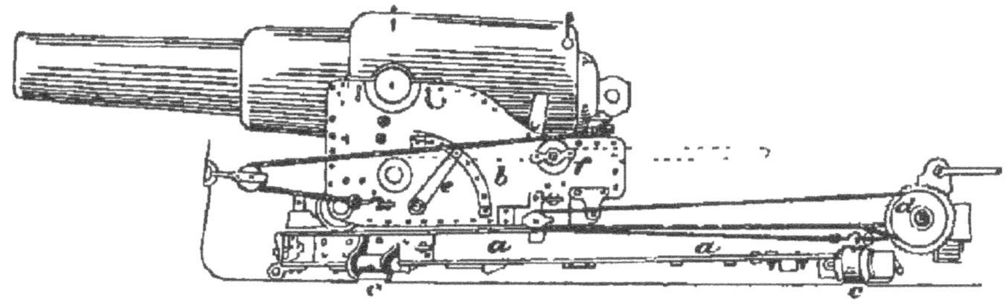

FIG. 30.—IRON CARRIAGE AND SLIDE FOR 7-INCH AND 9-INCH GUNS.

a, a. Slide. *b.* Carriage. *c.* Rollers of Slide
d. Running-in Winch. *e.* Adjusting Lever for Compressor.
f. Elevating Drum and Lever.

strong firm at Elswick, and is known as the Elswick plate compressor for checking the recoil of the gun. In the old wooden truck-carriages the carriage was free to recoil as far as the breeching of the gun permitted, which was generally sufficiently far to allow of the gun being loaded without having to run it further in; breechings are still retained for the heavy muzzle-loading guns mounted on the broadside, but simply as a precautionary measure in case of any failure of the compressor. When the first rifled guns were introduced, it became necessary however

[1] Royal Carriage Department.

to introduce some form of appliance for controlling the recoil, which should be automatic, but by which the gun could be kept under control when running in or out, without increasing the labour of these operations; it was also necessary that the shock of the recoil should be gradually absorbed, so as to avoid any sudden strain being brought on the mounting, and, finally, it had to be capable of adjustment.

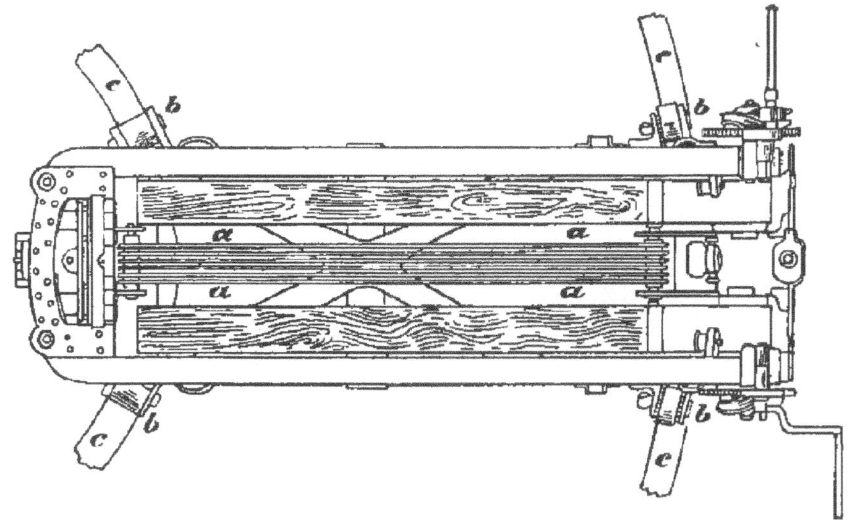

FIG. 31.

a, a a, a. Compressor Bars. *b, b, b, b.* Rollers of Slide. *c, c, c, c.* Racers.

All these essentials are combined in the Elswick compressor. Its action is frictional, and it has answered its purpose on the whole very well, although certain modifications were introduced by Rear-Admiral Scott, when he applied it to his carriages and slides, which were adopted for all guns above 9-inch on the broadside and in the earlier turret mountings. It may be described as follows, when fitted to the 7 and 9-inch mountings. Six long thin flat iron bars, *x x x* (fig. 32), some six inches in depth, are fixed centrally between the frames of the slide, which are attached to it at their inner and outer ends by bolts running through them and through brackets fixed to the front and rear transoms of the slide, the plates being allowed room for a slight lateral movement along the pins. Short plates, *z z z*, of the same depth are hung to the bottom of the carriage, so that each slide bar is between two carriage plates (fig. 32), and the whole can be compressed tightly together by means of short levers, *p p′*, called rocking levers, which act on the two outside plates in the carriage; these levers are connected with the compressor and adjusting shafts respectively; the compressor shaft, *c*, runs through the right bracket of the carriage, and has a lever, *y*, on its outer end; it is fitted with a left-handed screw-thread on which works a nut, *b*, at the head of the right rocking lever *a*, and with a right-handed screw-thread at its inner end, on which works the cylinder *b′* at the head of the left rocking lever *a′*; the motion of the shaft *e*, when the lever *y* is moved, causes the heads, *a a′*, of the rocking levers to move away from or towards one another, the levers turning about the pins *o o′*, and thus pressing the ends *p p′* together against the plates, or separating them; the shaft, *c′*, is called the adjusting shaft, and runs through the left bracket of the carriage, having a lever, *y′*, on its outer end; the inner end is feathered into the cylindrical part of the adjusting cylinder *b′*; the motion of the shaft *c′*, when the

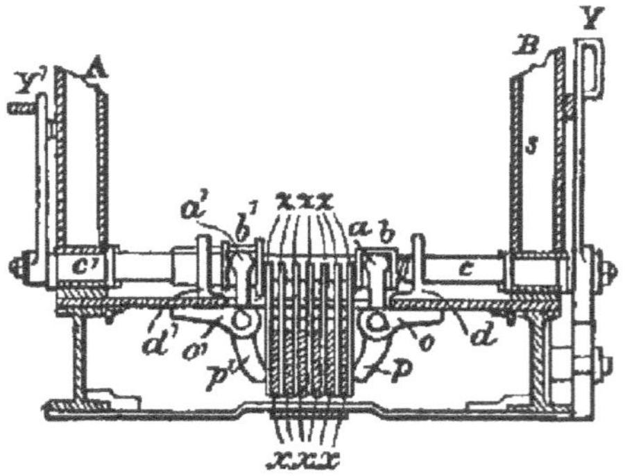

FIG. 32.—CROSS SECTION OF CARRIAGE SHOWING ELSWICK PLATE COMPRESSOR.

lever, z', is moved, causes the cylinder, b', to move either right or left, thus moving the head of the left rocking lever in its turn to the right or left, and so rendering it possible to put less or more compression on the bars or plates by means of the shaft *c*. After the gun is run out, the whole of the bars are compressed tightly together by the levers, so that considerable friction is caused on recoil, which is continued until the force of the recoil is expended. The compressor is made automatic by means of a

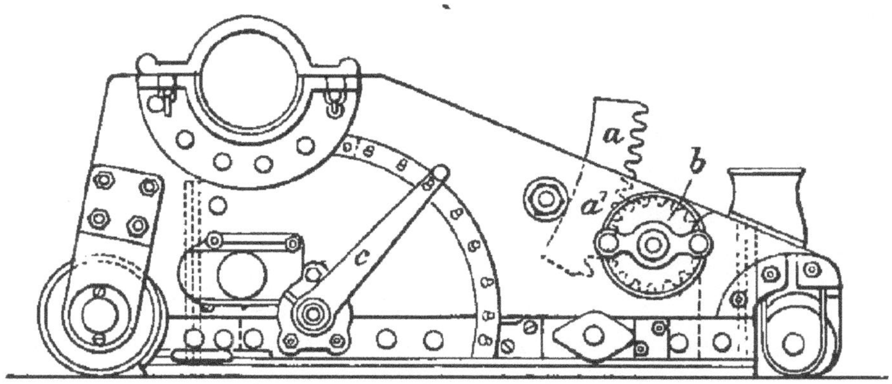

FIG. 33.—IRON CARRIAGE FOR 7-INCH GUN.

a. Elevating Arc attached to Gun.
a'. Tracing of Arc inside Bracket. *b*. Elevating Drum.
c. Compressor Adjusting Lever.

"tripper," which forces down the compressor lever as the gun recoils, if it has not been placed properly in position before firing.

Another improvement over the early wooden slides was the fitting of gun-metal rollers to the fore and rear ends of the slide, which, running on the racers, made the training of a 7-ton gun, by means of rope tackles hooked from the rear of the slide to the ship's side, a matter of comparative ease to a small gun's crew of seven or eight men, who would only have been able to work a much smaller gun, mounted on the old truck carriage, with a great amount of labour. In the case of slides for the heavier nature of guns above the 7-ton, the rear rollers are grooved to run upon ribs on the racer, while the training is effected, not by tackles, but by means of a training shaft, having at its lower end a toothed wheel, which works in a racer cogged for the purpose. This training shaft lies centred in the slide, and is worked by means of pinions and winch-handles at each side of rear extremity of slide. The racer is marked in degrees, so that by means of pointers on each side, the slide can be trained with great accuracy, when the guns are laid for an object by order from the officer stationed at the director in the conning-tower, or on the upper deck.

The guns are elevated by means of what is called the capstan-head elevating gear, which is fitted on each bracket of the carriage, and the gun is moved by means of an elevating arc pivoted each side to the gun, which on its rear edge has teeth, which gear in the teeth of the pinion at end of a spindle, and is kept in its place in gear with the pinion, and prevented from moving from the side of the carriage by a friction roller secured to carriage, and working in a groove on fore edge of arc (fig. 33). The spindles of the friction roller and pinion pass through sides of carriage, the former being secured by a metal nut, while on the latter is fixed the capstan-head and a clamp. The capstan-head has holes in it for the iron-pointed lever by which it is turned. When thus moved it turns the spindle, as it has feathers on it, which enter slots in the latter, and the spindle by its pinion works the arc, thus raising or depressing the gun as required. The clamp screws upon the spindle outside the capstan-head, so that when tightened upon the latter it clamps the pinion.

The Scott mounting, so called from its designer, Rear-Admiral Scott, is used for 10-inch 18-ton and 12-inch

25-ton guns, mounted on the broadside, and for the turret-guns in some of the earlier turret-ships. It consists of a lower and longer carriage than that of the R. C. D., which has just been described, and of a higher slide (fig. 34). The compressor is a modification of the Elswick, and is known as the "bow compressor," the action of the compressors in checking the recoil of the gun being assisted by one or two hydraulic buffers. The bow compressor (fig. 35) serves to control the gun and carriage in

FIG. 34.—18-TON GUN ON SCOTT BROADSIDE MOUNTING.

running in and out, as does the Elswick, but instead of being fitted in the centre of carriage and slide, it is constructed to act on each side of the mounting, and comprises on each side a bow or cramp, *a*, which passes through the bracket, and is pivoted near the centre on a pin, *b*, in a metal bearing, bolted to the outside of the bracket. The outer part of the cramp is fitted to take an adjusting screw, *c*, and wheel, *d*. The wheel is notched to receive a pawl, *e*, which keeps it in any required position; to the inner arm of the bow a plate, *f*, is hinged, and suspended from the carriage on the outside are three plates, *g g*, tapered downwards; attached to the slide are two long bars, *h h*, tapered upwards and lying between the plates *g g*; the sides of the slide are filled in with wood strips acting as additional bars; the revolution of the adjusting wheel, *d*, tightens the adjusting screw against the outside plate, and compresses bars and plates together, at the same time causes the inner arm of the bow to force the plate, *f*, against the wood strip on the inside of the slide. When the gun is put on the

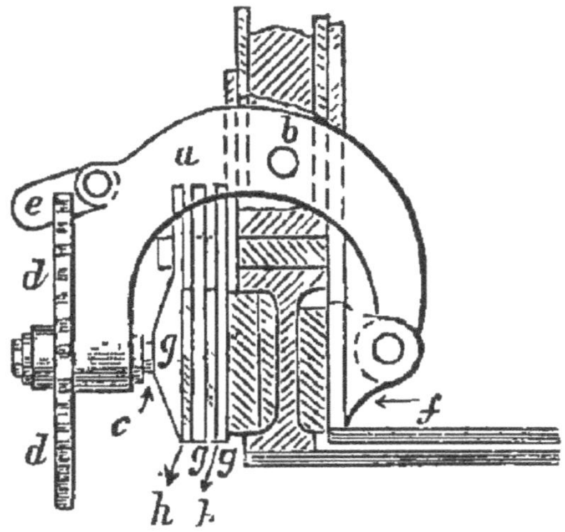

FIG. 35.—THE BOW COMPRESSOR, INVENTED BY REAR-ADMIRAL SCOTT, R.N.

rollers for running in or out, owing to the tapering of the plates and bars, the compression is removed. The Elswick and Scott compressors were later, in most heavy guns, either supplemented or entirely replaced by the hydraulic buffer, which is in use at the present day in nearly every slide in some form or other.

It consists in principle of a long cylinder attached to the slide, in which works a piston with a piston rod attached to the carriage. In the early patterns holes were bored in the piston. The buffer was filled with a certain quantity of

Rangoon or other oil, and the carriage running back on the slide during recoil forced the liquid through the holes in the piston, and the resistance thus produced checked the recoil of the carriage.

Buffers are termed either "tension" or "compression," according as the piston rods were pulled out or pushed in on recoil. The newer patterns have an arrangement by which the size of the apertures in the piston can be varied to suit varying charges, and also to act more gradually on the carriage during recoil. This is done, in some cases, by having projecting ribs inside the cylinder, along which slides a disc with deeper grooves in it; this disc is attached to the piston head. By turning the piston rod round, the clearance between the rib and the groove is altered, and so the size of the aperture is adjusted.

Other buffers have grooves inside the cylinder, of varying depth, so that the aperture becomes smaller as the rod advances and the velocity of recoil diminishes. Hence the resistance is kept nearly uniform during recoil.

Krupp's buffers, as well as some in our naval service, are entirely filled, and the apertures are valves, kept shut by a spring; consequently they hold the carriage in a seaway with the valves closed. On firing, the pressures are sufficient to open the valves, and the recoil is checked.

It is unnecessary to dwell at any further length on the mountings of the heavy muzzle-loading guns in the turrets of the earlier turret-ships, but in fig. 36 is shown a 25-ton gun as mounted in H.M.S. *Glatton*, the first coast defence turret-ship built and completed in 1871. But although the guns were worked by hand, steam training gear had to be provided for training the turrets. The *Thunderer* in 1875 was the first ship in which hydraulic machinery was fitted for working the guns; but in her case it was experimental, the fore-turret only being so fitted, the after-turret having the usual hand-worked guns. Her installation, however, proved so successful, that every battleship of later date has been provided with gun-mountings worked on the same principle, but much improved and extended. As this chapter refers solely to the muzzle-loading gun, and there are only three battleships of any importance, viz., the *Ajax*, *Agamemnon*, and *Inflexible*, which carry muzzle-loading guns heavy enough to require hydraulic mechanism, for the *Dreadnought* is to

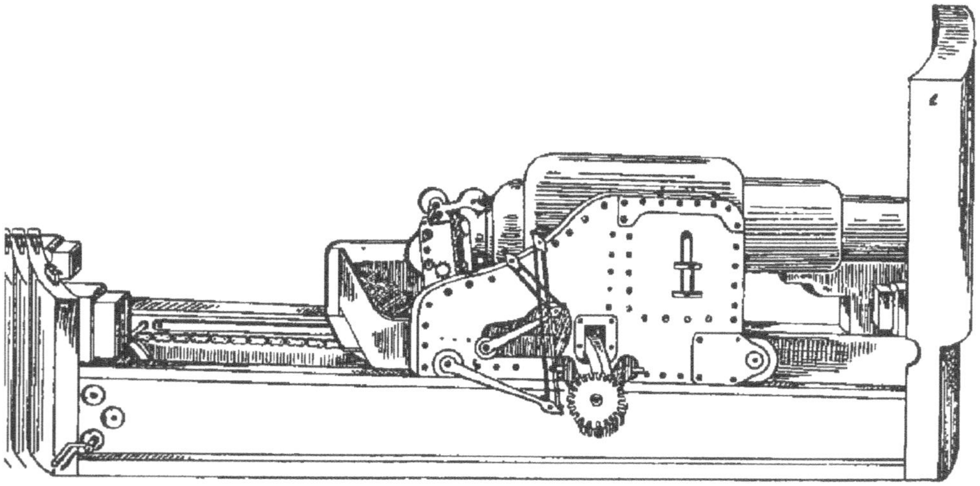

FIG. 36.—25-TON MUZZLE-LOADING GUN, AS MOUNTED IN THE TURRET OF H.M.S. "GLATTON."

have her muzzle-loaders replaced by breech-loaders, as has been done to her sister ships, the *Thunderer* and *Devastation*, no attempt will be made here to give any details of the hydraulic system, as applied to those ships. The guns are now mounted differently, they have no trunnions, the heavy muzzle-loaders of the three ships just mentioned were the last trunnion guns made. While trunnions existed large ports had to be made in the turrets in order to obtain the necessary elevation and depression, for although a complicated arrangement, involving great labour, was provided in a few of the earlier turret-ships, it was soon given up, and the large ports were accepted as a necessary evil. In the *Inflexible*, owing to the length of the guns, the loading had to be done outside the turret, the muzzle of the guns being depressed to meet the charge and projectile, which were entered and rammed home by the hydraulic rammer through an opening on the inner side of the glacis protecting the base of the turret.

CHAPTER IV.

GENERAL REMARKS ON BREECH-LOADING RIFLED GUNS, FROM THEIR RE-INTRODUCTION IN 1879.

The failure of the first Armstrong breech-loading guns, and the consequent adoption and introduction into the service of the muzzle-loading rifled guns, which have been described in a previous chapter, undoubtedly threw us back, as compared with France and Germany, some fifteen years in the art of gun construction; for it was not until the year 1879 that various causes, amongst them the great increase in the thickness of armour, and the corresponding necessity of obtaining guns giving higher velocities and increased penetrating power, combined to force the authorities to take up seriously the question of arriving at some satisfactory system of breech-loading. It was fortunate for the country that during the transition period which followed we were at peace, and it is gratifying to know that our ships are now provided with guns, mountings, and loading arrangements, which, taken as a whole, are equal, and in some respects superior, to those in foreign navies. And it is to the Elswick firm and to Sir A. Noble, its practical working head for so many years past, that the country is indebted for the bulk of the improvements which, during the last fifteen years, have completely revolutionized our naval gunnery.

In 1878, Sir A. Noble and Professor Abel carried out a

long and most exhaustive series of experiments with slow-burning powders, and longer guns made at Elswick, which established the fact that large charges of slow-burning powder, with adequate length in the bore for combustion of the charge, give higher velocity to a projectile than a violent powder in a short gun containing a similar weight of metal, the result being that much higher velocities have since been obtained, not only in England, but in other countries. It was at that time that Lord Armstrong again urged upon the Government the advantages of breech-loading in connection with long guns, and the importance of the results obtained at Elswick during the experiments; and in order that the Admiralty and War Office might satisfy themselves, 6-inch and 8-inch guns of new design were lent by the firm. The result of the further trials was, that guns were adopted giving muzzle velocities of 2,000 feet-seconds, whereas up to that time the highest muzzle-velocity afforded by any gun was 1,600 feet-seconds.[1] Large-grained powders were used during the experiments, and in order to obtain the full advantages from their use, and to secure the satisfactory ballistic results, it was necessary to use longer guns than heretofore. The action of the slower-burning powders is more regular and under control. A charge of small-grain powder is burnt almost directly it is ignited, as each grain takes an infinitesimally small time to consume; but when the size of the grain is materially increased, each grain will occupy some time in being consumed, and during the whole of this time it is giving off gas; thus, the continued generation of gas maintains pressure on the projectile during the whole time it is travelling along the bore. It is thus evident that to obtain the full advantage from slow-burning powders, long guns

[1] Lloyd and Hadcock, "Artillery, its Progress and Present Position."

are a necessity, as with them more work is done by the powder on the projectile. When the large-grained powder is used from the obsolete short muzzle-loading guns, still mounted in our older battleships, the projectile leaves the gun before the large grains of powder are completely consumed, with the result that the latter are blown out still burning; a fact which has been proved continually by the scoring of the deck over which the turret-guns have been fired.

Experiment further proved that a greater amount of large-grained powder can be burned without increase to the maximum pressure if the charge is "air-spaced," that is, by making the capacity of the powder-chamber greater than the volume of the actual charge. In this way the mean pressure on the projectile, while travelling along the bore, can be increased: the result is higher velocity without undue increase of pressure. With the increase of the charges, however, it became necessary to place a limit on the length of the cartridge, as long cartridges are most inconvenient for storage, and, by extending too far down the bore when rammed home, they place the projectile in an unfavourable position, the length of the bore being also practically shortened To get over these difficulties, it was determined to enlarge the powder chambers by increasing the diameter, as shorter cartridges could then be employed, while the requisite thickness of metal could be easily added externally. With muzzle-loading guns this enlargement, or chambering, as it is called, could not be carried out to any great extent, still it was applied with good results to the 38-ton 12·5-inch, the 80-ton guns, and the 13-pounder field gun; in the former it was found that with a chambered gun a charge of 210 lb. prismatic powder could be fired, giving a velocity of 1,575 feet-seconds, whereas with the old form of chamber, these guns fired 160 lb. P 2 powder, with a velocity of 1,442 feet-seconds. With breech-loading guns there is no difficulty, provided that the entrance to the chamber is as large as the cylindrical part; the chamber then may have any capacity, and the cartridge can be made to suit its dimensions. The necessity of having longer guns in itself tended to force breech-loaders on the authorities, as length was a serious objection to muzzle-loaders, involving, in the case of large turret-guns, the loading arrangements being fitted outside the turret. Where the guns were loaded by manual labour, as in the guns of moderate size, increase in the length involved great manual exertion, while in breech-loading guns the distance for ramming the charge can never be great; ordinary power will suffice, and less space is required, as the gun need not recoil so as to bring the muzzle within the port. Another advantage which breech-loading guns have over muzzle-loaders is, that where guns are mounted on the broadside, the loading numbers at a breech-loader are necessarily not nearly so exposed as at a muzzle-loader, where the men are extremely liable to be picked off through the ports. Finally, another powerful reason in favour of breech-loading guns is the possibility of repairing them by renewing the inner tube (or liner as it is called) as erosion takes place in the bore. With the heavy charges that are necessary for high ballistic power, the scoring in the bore seriously affects the life of the gun. Muzzle-loading guns can certainly be repaired with a new barrel, but this generally involves a new tube over the chase as well.

One of the objections urged against breech-loaders was, that in opening the breech after firing, a large quantity of smoke would fill the turret or battery, but, as a matter of fact, far less smoke finds its way inboard, as the greater part escapes by the muzzle which is outside the port; while with a muzzle-loading gun the recoil into the loading position brings the muzzle, with the smoke still issuing from it, inside. The great danger incurred at first with breech-loading guns was, that they might be fired without the breech being properly closed. This, however, has been rendered almost, if not quite, impossible by automatic safety arrangements attached to the breech action itself, which prevent the guns being fired until the breech is properly closed.

It was only in 1881 that, after exhaustive trials, the manufacture of the new breech-loading ordnance fairly commenced; and, from that date on, the immense strides that have been made in the method of construction of the guns, no less than in their greater power, their increased rapidity of fire, and the efficiency of the mountings, are striking testimonies to the mechanical and engineering genius of the present day. It may be said that there are a few points of resemblance which run through the whole class of breech-loading guns. One of the principal features is length, which is required to allow the full development of power from the large charges of slow-burning powder now used; the length of bore varies with different natures, breech-loading field guns being about 25 calibres, the heavy guns 30, and quick-firing guns 40 to 45. There is a swell at the muzzle to give strength to a part most exposed to an enemy's fire, except in the case of light quick-firing guns; and greater strength also here is now required on account of the muzzle pressures being much higher than formerly. The powder chambers are greatly enlarged, cylindrical in form at the breech to admit a cartridge of the largest diameter possible. The rifling groove is of the polygroove hook section in all but the quick-firing and certain guns supplied by Elswick; the dimensions of the groove at one time were the same for all natures of guns, but the depth now varies slightly with size; in field guns it is ·04, in the medium natures ·05, and in the heaviest guns ·06. The number of grooves corresponds with four times the calibre in inches in all but the quick-firing and certain guns supplied by Elswick, which have rather more, but 12-pounders and 4-inch breech-loading guns are rifled on the principle of six grooves to every inch of calibre. Originally the twists of rifling in all R.G.F. guns increased from a small amount at the breech to a maximum about halfway down the bore, and from there to the muzzle it was uniform at the maximum pitch; but since 1888 all guns of 6-in. calibre and upwards have been rifled with an uniformly increasing twist from one turn in 60 calibres at the breech to one turn in 30 calibres at the muzzle.

One of the principal, if not the principal point to be decided when it was determined to return to breech-loading guns, was an efficient method of closing the breech; for as has been already pointed out, the great fault in the original Armstrong breech-loaders lay in the fact that the gun could be fired without the breech being properly closed. The choice lay between the wedge system of Krupp and that adopted in France, known as the interrupted screw, and the latter was selected as combining the advantages of other systems without their drawbacks. This system is applied to all breech-loading guns in the service, except the smaller natures of quick-firing guns. It consists of a solid breech-block of steel, furnished with a screw-thread of the requisite strength and pitch, while the gun is prepared with a similar female screw to receive it; the surface of the block is then divided longitudinally into six or eight equal parts, and the screw-thread is entirely removed from alternate portions by a planing machine; in the gun alternate portions are slotted away, but the parts which correspond with the screwed portions on the block are left in relief, and those which are opposite to the planed part are cut away. When mutually prepared in this manner, the block can be pushed into the gun or drawn out with direct motion, while a turn of one-sixth or one-eighth of a circle (after being pushed in) is sufficient to bring all the screw-threads into gear. The bearing surface and strength of the screw is obviously reduced by one-half, but the requisite strength is easily provided by regulating the length of the block, as such length cannot appreciably affect the time occupied in opening or closing the breech. Although additional mechanism is supplied for working the breech-blocks of the heavier natures of ordnance, this extra leverage or power does not alter the principle of closing the breech.

To open the breech-screw in this system four motions are necessary. First, to unlock it (in the larger guns, with Stanhope lever, by raising the lever); second, to unscrew it, a fraction of a turn corresponding to the number of interruptions in the screw; third, to draw it directly to the rear on to the carrier; fourth, to swing it round to the side of the gun. To close the breech the above motions are performed in the opposite direction and order.

Although the firing-gear has been frequently altered, and is now in ordinary breech-loaders (not quick-firing guns), in the opinion of many competent authorities, still far too complicated to be satisfactory, it has yet been always so arranged that the gun cannot be fired until the breech is closed and locked, and no accident from failure of these precautionary arrangements has yet been recorded with English breech-loading guns.[1]

Another important point to be secured was that the breech-block should have complete obturation; that is, that there should be no escape of gas to the rear, through the

[1] Lloyd and Hadcock, "Artillery, its Progress," etc.

breech-block not fitting closely. In the first breech-loading guns introduced into the service, obturation was obtained by the Elswick cup, as it was termed, but which in some respects proved to be unsatisfactory, so the system known by the name of De Bange (so called after the French general who invented it), which had been used for some years in France, was tried, and after prolonged experiments, of which the results were most satisfactory, was finally

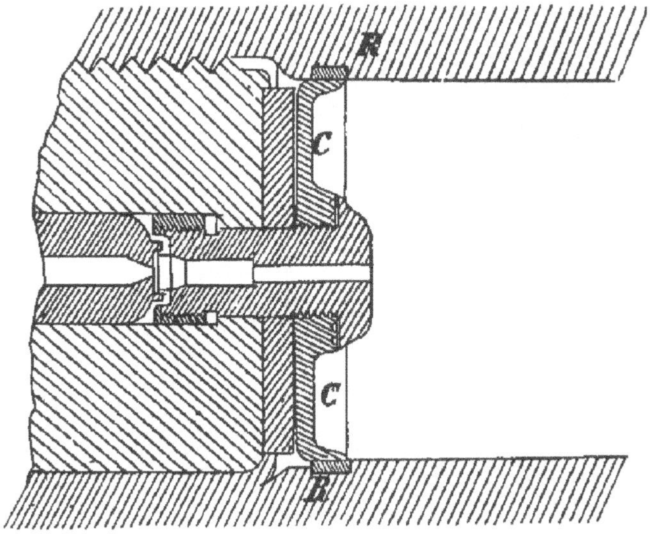

FIG. 37.—THE "CUP" OBTURATOR.

adopted for all breech-loading guns in 1882, except those already fitted with the Elswick cup.

The "cup" system (fig. 37) consists of a shallow steel disc, C, cupped in front and bolted axially to the face of the breech-block by means of a spindle and nut in the centre. The back of the cup is flat, while the face of the breech-block is made slightly convex. A copper ring, R, is let into the gun, behind a step in the chamber, for the edge of the cup to take against when set home into contact by the travel of the breech-screw. When the gun is fired, the pressure of the gas drives back the outer portion of the cup upon the rounded face of the breech-block, so that the rim is forced out against the ring, and, bedding itself in the copper, prevents all escape of gas. The central bolt that kept the cup in place served also for the vent. With service the copper ring got worn and great care had also to be taken to keep both it and the cup perfectly clean, for if any dirt got between the ring and the edge of the cup, close contact was prevented, and there was consequently an escape of gas which damaged both ring and cup. It was easy to replace a damaged cup, but the renewal of the ring required skilled artificers and special tools.

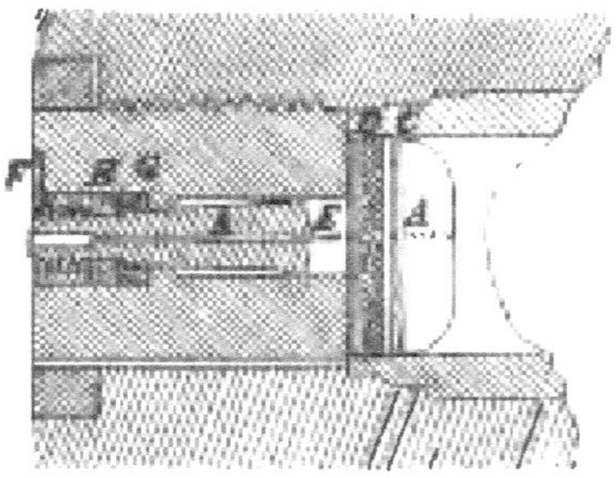

FIG. 38.—THE DE BANGE PAD AND OBTURATOR.

The De Bange pad is a cushion made up of asbestos soaked in tallow, reduced to shape by pressure in a hydraulic machine and inclosed in a strong canvas cover. The obturator consists of a mushroom-headed steel spindle, A (fig. 38), which is in immediate contact with the powder charge, the pad, B, two metal protecting plates or discs, made of tin strengthened by outer rings of steel, C and D, and a steel adjusting disc, E; the spindle, A, passes through the centre of the breech-screw, in which, when secured, it is still free to move slightly to the rear, so that, when the gun is fired, the pressure crushing the pad between the mushroom-head and the breech-screw, compresses it with respect to its thickness and correspondingly expands it in diameter, the result being, as experience has proved, that no gas can escape between the edge of the pad and the pin, or round the central hole in the pad and the spindle. After the pressure is removed, owing to its elasticity the pad recovers its former shape, and on the breech-screw being unlocked the obturator can be withdrawn by a straight pull. In 6-inch Mark III. guns, and upwards, the whole obturator is held tightly back to the breech-screw by a steel spiral spring, G, encircling the spindle; behind the spring comes a steel washer, and then a nut, H, is screwed on; this arrangement lessens the jar on firing. This system of obturation is extremely simple and has proved perfectly effective in guns of every size; it involves but little care in use, and there is nothing in the gun which can ever require repair. These pads will last a long time, being almost indestructible, except perhaps from the wear and tear of opening and closing the breech, and a new one can easily be substituted when necessary by any one who has seen the operation once performed. Some pads have been known to last thousands of rounds, but if the firing is rapid the pad may get slightly compressed, which can be remedied by inserting more adjusting discs; or it may get softened by heat, when it should be changed and thrown into cold water for a time, when it will soon be restored to good condition again; but whatever the size of the gun it need never be thrown out of action for want of repair, as in the case of the cup system of obturation, an advantage the importance of which cannot be over-estimated.

In both systems of obturation it is essential that the pad or cup should be attached to the breech-screw by means of a spindle and not rigidly fixed, for after each round there is a tendency on the part of the obturator to stick to the side of the chamber, which would render the work of releasing the screw an operation of difficulty, but by means of the spindle attachment, the breech-screw is free to turn independently, while its removal from the gun is then easily effected by a direct pull to the rear.

The first of the new breech-loading guns introduced into the service in 1880 were 6-inch guns made by the Elswick firm; these were constructed of steel inner tubes or barrels, strengthened by wrought-iron hoops. The next pattern guns were made at the Royal Gun Factory, Woolwich, and were designed on the following principles: first, the interrupted screw system for closing the breech, which has been already described; secondly, a continuation of the latest method of construction of rifle muzzle-loading guns, viz., a steel barrel strengthened by wrought-iron coils, with the trunnion piece welded to the breech coil.

In 1881, on the recommendation of Colonel Maitland, it was decided to make guns entirely of steel, and plans were drawn out; but at Elswick, where some of the new type were to be constructed, objection was taken to the designs, and this led to a fresh investigation by the Ordnance Committee, when the following report was adopted:—(1) That the superiority of steel over wrought iron was so marked that the latter should be entirely abandoned; and, (2) that coiled steel, though superior to coiled iron, was inferior to cast steel forged out into hoops. The committee therefore recommended that guns should be constructed entirely of steel. New designs were accordingly prepared, into which various modifications and improvements, however, had to be introduced, as the first designs were found to have several serious defects. The barrels increased in both length and diameter, until what was at first a steel lining became a steel body or core for the gun. Steel coils were first used for the exterior cylinders, and were made from bars of very mild steel coiled and welded like iron; these, however, were soon superseded by solid forgings of steel; the steel being cast, forged and mandrilled out into hoops, which are shrunk over a barrel of steel, itself cast and forged from a very thick ingot to the required dimensions and length. Hoops of cast steel supply the strongest material known except steel wire. With the use of steel hoops a new departure was taken in the method of building a gun; for, in the first place, it was no longer possible to obtain longitudinal strength, as in the case of wrought iron, by welding the cylinders end to end; and, secondly, the degree of shrinking most suitable for steel proved to be very different from that which had been formerly given to iron.

The question of shrinkage was worked out by experimenting with unserviceable guns, and it was shown that the permanent extension of a steel hoop may be two or three times as great as that which was given to a wrought iron coil. Greater heat is required for the temporary expansion of a steel hoop than was formerly required for wrought iron coils, and as heat is liable to affect the quality of the steel, great care has to be exercised in the operation of shrinking; the heat required has been found to be about 600° Fahrenheit, which is not sufficiently high to injure the steel, for in tempering, or oil-hardening (the more correct term) the metal, more than double that heat is employed. Steel is, what is called, a carburet of iron, and although the proportion of iron and carbon may vary to a great extent, it is absolutely distinct from cast iron, all the carbon in steel being chemically combined with the iron. Wrought iron, steel, and cast iron will all be hardened, more or less, if heated and then suddenly cooled; the effect on wrought iron is very slight, and on cast iron it is unattended with any springiness, but it is a property peculiar to steel to receive hardness and elasticity from chilling, although the degree in which these qualities are acquired by the treatment varies very much according to the amount of carbon present. For instance, what is termed mild steel contains no more than from ·1 to ·2 per cent. of carbon, and it cannot be appreciably hardened; whereas high steel, containing perhaps 1·5 per cent., is quite altered by sudden chilling, and becomes so hard that tools made from it can only be sharpened on a stone.

Hardening increases the breaking strength of steel; thus a steel containing from ·23 to ·3 per cent. of carbon might before hardening break at from 30 to 35 tons; but after that process at not less than from 40 to 46 tons. To obtain a very high breaking strength steel must contain a large proportion of carbon, and must be hardened.

What is usually understood by soft steel will not have a breaking strength of more than about 26 tons, a condition explained by the fact that the high elongation required in soft steel is inconsistent with high breaking strength and high limit of elasticity. If the two latter are increased by the addition of carbon and by hardening, the elongation must correspondingly decrease, and *vice versâ*. The steel must, therefore, be manufactured and hardened in such a manner as to develop the particular qualities required for the purposes it is to fulfil; for gun steel, although high elasticity and breaking strength are sought after, it is also necessary to have good elongation, so a compromise retaining all three qualities in good proportion must, therefore, be aimed at. To temper steel it has first to be hardened, and after that process again heated moderately.

It has already been stated that the barrel at first extended from muzzle to breech, and that the breech-screw geared into the barrel; outer hoops were then shrunk on until the transverse strength was complete, shoulders being duly arranged to prevent any displacement of the barrel from the trunnion ring, which holds the gun in its carriage. It was soon found, however, that this mode of construction brought too great a strain on the metal of the gun in proximity with the face of the screw, and there was in addition a tendency to shear the threads of the screw, because the chamber part of the A tube expanded on firing, while the rear part did not do so, and thus the threads had an unequal bearing. One of the first improvements, therefore, in the new guns was to make a breech-piece or jacket into which the breech-screw should gear in order to relieve the barrel. The inner diameter of the breech-piece at the breech end was contracted by steps corresponding to steps on the barrel (fig. 39), and the barrel made to end on a level with the face of the screw, the obturator alone being permitted to enter the bore, so that all cross strain was abolished.

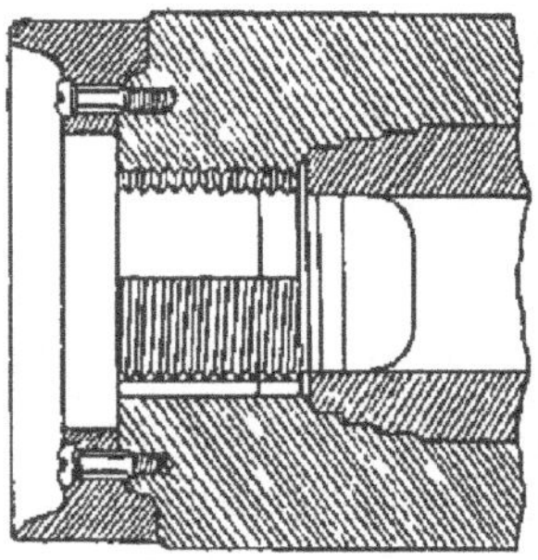

FIG. 39.

The breech-piece in the heavier guns is a cylinder of steel, made in the same manner as the barrel itself, differing only in length and diameter; there are various ways of linking this part of the gun to the trunnions, while it may be supported by exterior hoops to complete the requisite strength transversely. In the lighter nature of ordnance the trunnions are forged on the ingot which forms this part of the gun, and it is then known by the name of a "jacket."

Colonel Maitland, R.A., introduced an improvement of great importance by uniting all the principal parts of a gun together by a system of bayonet joints. The jacket or breech tube in this case is prepared at one end with a row of projections on the inner surface, while the body of the gun is prepared in like manner with a corresponding row of projections without; in the operation of shrinking (*i.e.*, in building up the gun), the projections on the one part pass between those on the other; the outside portion is then turned round in position until they engage one another. All the intervals are afterwards filled up by wedges driven in very tightly; one wedge would be sufficient to key up the parts, but by placing them in every space, and by giving them a uniform taper of 1 in 100 (heavy guns 1 in 200), continuity of metal is preserved to meet any stress at that part in a circumferential direction, while the strain is distributed more evenly through the material, and any forward movement of the barrel rendered impossible.

This system of connecting the hoops to each other or to the barrel, however, does not meet with favour at Elswick, where it is thought to be open to many serious objections. Commander Lloyd considers: (1) that it is difficult to insure that the joint at one end of the hoop and the shoulder at the other are both in contact; or, if in contact, that sufficient length has been allowed not to cause severe longitudinal strain; (2) that the covering hoop is seriously weakened by the lugs or projections, and it is, therefore, necessary to supply extra strength to the gun at the cost of additional weight; (3) that the construction is very expensive; and, lastly, that it is impossible to control the strain put upon the hoops when the keys are being driven in.

A serious error in the early construction was the non-hooping the barrel along its chase, or forward part. A very serious accident on board the *Collingwood* in 1886 (nobody, fortunately, being hurt), when one of her 43-ton guns was destroyed, exposed the weakness of the chase, and showed the necessity of strengthening the guns towards the muzzle. Only a reduced charge was being fired from the gun in question, which had already been proved in the usual severe manner, when the whole chase, or unhooped portion, was blown off, leaving the rear portion undisturbed.

The committee of investigation reported that the accident

was due "to a combination" of the following causes: (1) To a want of uniformity in the metal of which the chase was composed. (2) To the absence of annealing after forging and oil hardening, which treatment would have mitigated any internal strain set up by these processes. (3) That such strains were probably intensified by the preliminary and proof rounds, and this defect was developed during the interval of time between proof and accident. (4) To the foregoing unfavourable conditions not having been counteracted by the hooping of the chase.

One of the results of the accident was the decision to strengthen the chase of all guns that had not been so built, and this was done by shrinking on a number of chase hoops, and more recently one or two tubes instead of a number of short hoops.

A further provision for strength has been made by building up the heavier guns in a fewer number of parts, and such improvements made in the designs that the external parts have much more share in resisting the longitudinal strain on firing.

There are consequently amongst breech-loading guns a great variety of constructions, and it is not necessary to recapitulate them all, but merely to give the latest improved patterns known as: (1) the fourth R. G. F. (Royal Gun Factory) system, embodying a breech-piece to take the breech-screw, and a few exterior tubes of great length, shrunk on with interrupted projections, so that all parts are united together in a manner which enables them to share the strain in any direction.

(2.) The E. O. C. (Elswick Ordnance Company) construction, as applied to the 16·25-inch 110-ton gun, viz., a barrel and breech-piece with a large number of small exterior hoops extending from muzzle to breech.

(3.) The latest R. G. F. system, on which the 9·2-inch,

Mark VII., and Marks IV. of 10-inch and 15·5-inch guns have been constructed. The novelty of construction in these guns consists in, that the principal parts are put on, in building up, from the muzzle (instead of from the breech, as was done originally), being secured in position by steel bushes screwed in at the breech end; in the case of the breech-piece, the inner surface of the bush is screw-threaded to take the breech-screw.

But the last three years have seen a fresh development in gun construction here in England, which is superseding completely even the most improved patterns of steel guns built up in successive layers of steel, viz., the wire-built gun. The system of winding on wire has been advocated for many years, but it is only lately, after long years of experiment, that it has been finally adopted by the Woolwich authorities. A description of the wire-built gun will be deferred until another chapter, as must also that important development and improvement in the breech-opening mechanism of the lighter natures of guns, which, with improved ammunition, has resulted in the universal adoption of what is known in these days as quick-firing ordnance.

CHAPTER V.

REMARKS ON THE CONSTRUCTION OF BREECH-LOADING GUNS (CONTINUED).

THE design of a gun being approved, the exact dimensions of the various parts as intended for the completed gun are fixed, and the manufacture can be commenced, the casting of the steel ingots being the first step in the process.

Steel may be considered an alloy of iron with carbon, manganese, silicon, sulphur, and phosphorus. The two latter are extremely injurious to the metal, as they produce brittleness, and therefore they have to be got rid of as far as possible. All steel is made by some process of fusion, and in England the process adopted for gun steel is what is known as the Siemens-Martin, or open-hearth method. Swedish and Spanish pig-iron is alone employed now for making gun steel, as the ore is much purer than any other.

An extremely high temperature is required to melt the steel, which must not be in contact with the solid fuel, because if it were so, the proportion of carbon would be too large, and special measures have also to be taken to render the product as uniform as possible throughout the mass. The charge put into the open hearth consists of pig-iron and scrap steel in the proportion of about two of pig to one of scrap. It is put into the hearth and subjected to the full heat of the furnace for about three and a half hours, and when melted a little hematite ore is added; when, first,

manganese, then silicon, and lastly carbon, combine with the oxygen in the ore and pass away with the waste gases, the process producing a violent ebullition, termed boiling. When the carbon is eliminated to the desired point, and just before the charge is tapped, ferro-manganese is added, in the proportion of about ·5 to ·9 per cent.

It is entirely to the accuracy with which the carbon is eliminated, and the manganese added that the suitability of the steel is due. It is necessary to add manganese to the steel to make it suitable for forging, and it also to a certain extent counteracts the evil effects of sulphur, should any exist in the metal, while it forms with the silica a fluid slag which removes other impurities; it also acts like carbon in hardening the steel, but less energetically, and it lessens the ductility.

Carbon exists in steel in proportions varying from ·15 to 1·75 per cent; in steel as used for ordnance, usually from about ·24 to ·34 per cent. Increasing the amount of carbon increases the fusibility, hardness, and tenacity of the steel; but diminishes its ductility and malleability. The carbon gives rise to the important properties of hardening or softening by rapid or slow cooling; but the higher the amount of carbon, the more is the metal likely to be injuriously affected by the internal strain, set up by the hardening process, and the greater is the difficulty of the subsequent forging operations, which have to be done at a lower temperature as the proportion of carbon increases.

When the fusion is complete, the molten metal is run into ladles resembling large cauldrons, lined with brick, and fitted so that the liquid metal can be tapped from an orifice in the bottom. The charged ladle is then taken over the ingot mould, and the steel allowed to run into it. With care several furnaces may be arranged so that the steel in them all is ready for tapping at the same time. By this means ingots weighing over forty tons are frequently cast.

Almost all of the impurities, especially phosphorus, have a tendency to rise to a point near the head of an ingot (the final point of solidification) while it is in the liquid state. Every endeavour is made, either by the use of silicon, different methods of cooling, or by subjecting the fluid ingot to pressure, as practised by Sir J. Whitworth and Co., to make the casting as round and uniform as possible, but when cooled, only about the lower two-thirds are solid throughout, while the upper portion, which contains most of the impurities and gases which have been unable to escape, presents a spongy appearance and is chiefly waste. The head is therefore sawn off as soon as the ingot has cooled, and when this is done the steel is in readiness for the manufacture of a barrel or a hoop for a gun.

The ingot is next heated and forged, which is carried out by heating in large furnaces, and then, by means of either steam hammers or hydraulic presses, hammering or pressing the ingot down to the desired shape.

All ingots are cut much shorter and thicker than the dimensions required for the forging. This is not only more convenient for casting, but greater homogeneity of the metal is obtained; also the more forging work that is expended on an ingot in drawing it down, the better will be the quality of the steel. The forging of steel castings, however, requires great experience and skill, as particular care has to be taken that the steel is not overheated, nor should it be drawn down very much at each operation.

If the casting is intended for a long tube, it is alternately heated and forged out by degrees under a steam hammer (or by means of a hydraulic press) into a cylindrical shape, until its diameter is reduced to from about one to one and a half inches above that required for the finished tube. The test pieces were, up to a comparatively speaking short time ago, cut off the forging at this stage, and if found not up to the specification, which, as mentioned in a previous chapter, should be from thirty-four to forty-six tons breaking strength, twenty-one tons limit of elasticity, and 17 per cent. elongation in two inches, the forging was rejected. Under the existing regulations, however, the test pieces are now not cut off until after the forging has been hardened and annealed. The forging is next rough bored and turned, and is then annealed by being heated and allowed to cool very slowly, the object of this operation being to remove the internal strains produced by forging. Oil hardening, or tempering, as it is sometimes called, is the next process. The heating is done in a vertical furnace, wood, as being cleaner, being used for fuel. When heated to about 1,450° Fahrenheit (it varies from 1,350° to 1,600°, according to the hardness or mildness of the metal), the forging is lifted out of the furnace by a large overhead crane, and lowered into a deep tank of rape oil, which is at a temperature of 60°; it is allowed to cool gradually for some seven or eight hours, and is then lifted out and again annealed by being reheated in the furnace gradually to about 900° Fahrenheit, after which it is allowed to cool very slowly to the atmospheric temperature. Test pieces are now cut off. The only part of a heavy gun not oil hardened is the trunnion ring, as, from its irregular shape, the treatment would cause warping and internal strains.

The oil hardening and subsequent annealing are carried out because all steel of the quality used for guns, possesses the important property of becoming harder and stronger by more or less rapid cooling from a high temperature. The more carbon and manganese in the metal, and the lower the temperature, and the greater the power of absorbing heat in the cooling liquid employed, the greater the hardness produced. Water being a good conductor of heat, hardens more actively than oil; but the latter is considered the best for gun steel, as the more gradually the action takes place, the more thorough and even is the effect on large masses, and the less the risk of setting up dangerous internal strains. There is a great increase of elasticity and tenacity obtained by oil hardening, but it is gained at the expense of ductility, the power of elongating before fracture being considerably diminished, and there are other disadvantages, the most important being the internal strains which are produced, due to the varying rates of cooling of different parts of the mass, and which it seems impossible to prevent. It must be continually borne in mind, that for gun construction the steel must be free from internal strains. When these are present the steel is in a condition of unstable equilibrium, most dangerous when enormous forces have to be dealt with, as in a gun.

It is to remove as far as possible the internal strain that the steel is subjected to a second process of annealing, as has been described, after the oil hardening; the hardening effects are thereby considerably modified, the tenacity is lowered, but the elongation per cent. is rendered higher, as is the yield point.

If the testing shows that the steel complies with the specification, the forging is now second-turned and second-bored, and if intended for an A tube, it is subjected to a water proof, in order to detect any flaws; the ends are tightly closed, and water forced in from a hydraulic main up to a pressure of about three tons on the square inch. The hoops are meanwhile treated in the same way, and the different parts are then ready for heating and shrinking on. The amount of shrinking required depends upon the powder pressure, due to the charge for which the gun is designed, allowance being generally made for about double the pressure in order to provide a good factor of safety.

The outer surface of an inner part is very carefully turned to the required number of thousandths of an inch larger than the interior diameter of the part that is to be shrunk on to it; the inner surface of the latter having been previously bored to exact dimensions. Great care has to be taken that the external diameter of the inner hoop or barrel exceeds by exactly the right amount the interior diameter of the hoop to cover it. If this is not done, either there will be too much extension of the outer and compression of the inner hoop, or too little. In the first case the result will very likely be that the outer hoop will be strained beyond its elastic limit when the gun is fired, and that an extension above what is required will be caused, and then the internal hoop will be deprived of the support it ought to receive. The operation of "turning for shrinkage," as it is termed, requires most accurate and repeated gauging. The hoop or breech-piece, as the case may be, which is to be shrunk on, is placed under a crane and raised by a wood fire to a heat of about 600° Fahrenheit. It is then raised by the crane and lowered over the barrel, which has previously been placed breech downwards in a pit (as all the parts are now put on from the muzzle end), until a shoulder machined within them rests on a corresponding shoulder formed on the outside of the barrel.

When steel built-up guns were first introduced, short hoops were employed, but in these days of very long guns, *girder* strength, to support them from bending, is specially required, and long hoops will better provide this strength than short ones, so they have now been adopted.

Since heat expands the hoops longitudinally, as well as transversely, therefore, as the new hoop cools, it will contract in length. In order to avoid gaps between the hoops, the gun should be designed so that the shoulder on which the new hoop is to rest is not far from its lower end. One end of the hoop should always cool and grip before the other. To provide for this, a ring of burning gas-jets, or a highly heated ring of iron is put over the portion of the hoop which should be the last to grip; it will then remain expanded, and be free to move until the other portions of the hoop have gripped. If the two ends of the hoop were allowed to cool first, as they would if left to themselves, the middle portion of the hoop would cool in a state of tension. In order to insure that the hoop shall cool from its interior, a stream of water is made to play up the gun during each operation of building up.

Immense experience is required in performing all these details, for on them the strength and reliability of the gun depends: and any carelessness or ill-judged treatment is certain to be followed by failure and perhaps disaster.

The Woolwich method of connecting hoops to one another or to the barrel was described in the last chapter, but hoops can also be connected together by screwing. The hoop about to be put on is provided with the outer or female screw, and it is screwed up while hot. In the opinion of the Elswick authorities, no better system of joint exists, but they consider that the difficulties of practical application are considerable, and the result is that it becomes expensive. The hot hoop must be screwed up very quickly, for if it cools and grips before the operation is complete, it will almost certainly have to be cut off and destroyed. Suppose the screwing to be successful, the other end must be kept hot, so that it will be drawn down to its shoulder by the screwed end while the hoop cools. But here again the shrinking has to be allowed for very accurately, or the result may be that it will split, where it is weakened by the screw-cutting, when the gun is fired.

Another and apparently simple method of connection is by "serrations." With them the hoops are not weakened by being deeply cut, for the serrations are so shallow that those on the heated outer hoop pass over their fellows on the cool inner one. Construction is therefore very simple, although great care must be taken that the serrations shall be machined so accurately that they shall all grip practically at the same moment, otherwise it might happen that the longitudinal strain will come only on one.

After the building up of the gun has been completed, it is removed to the turnery, where it is placed in a lathe, finish-bored and chambered; then comes what is called "broaching," very fine boring in order to obtain as perfectly smooth a surface as possible. The next operation is rifling. For this the gun is placed horizontally in a machine; each grove is cut singly, the form of groove being given by the shape of the cutter. When one groove is finished the gun has to be turned on its axis to another position, and the next groove is cut in a similar manner. After rifling the breech is drilled to receive its fittings; these, such as axial vent, breech-screw, etc., are made from forgings of mild steel rough turned, oil hardened, etc.

The gun is next handed over for "proof." The bore is very accurately gauged to thousandths of an inch at regular intervals, and complete sets of gutta percha impressions are taken. Six rounds are then fired at the proof-butts with flat-headed bolts of equal weight to the service projectile; some of the charges are proof charges, *i.e.*, giving 25 per cent. higher pressure than the service charge, and some are ordinary service charges, the proportion of each varying with the nature of gun. After firing, the gun is again carefully examined, the results of gauging and impressions being compared with those previously taken ; if the alterations in dimensions are not abnormal, and if no new defects have developed, nor any slight mark perceptibly increased, the gun is passed as regards its firing proof. The impression of any defect is preserved for reference.

The gun having been passed is now returned for completion, the most important part of which consists in drilling the holes for the sight-sockets, and fitting the sights (except for the 12-inch guns and upwards, the sights for which are attached to the mounting). Other operations consist in cutting and engraving the various lines and marks on the gun; on the top of the breech is marked the angle at which the sight is inclined for drift, the broad arrow, the actual weight of the gun, the Royal monogram, and a small rectangular portion is filed level to form a clinometer plane. On the right side of the breech and muzzle small lines are cut to indicate the line of horizontal axis On the face of the muzzles, horizontal and vertical lines, the material of and register number of the A tube, also a II if the gun has Mark II rifling.

The trunnions are marked as follows :

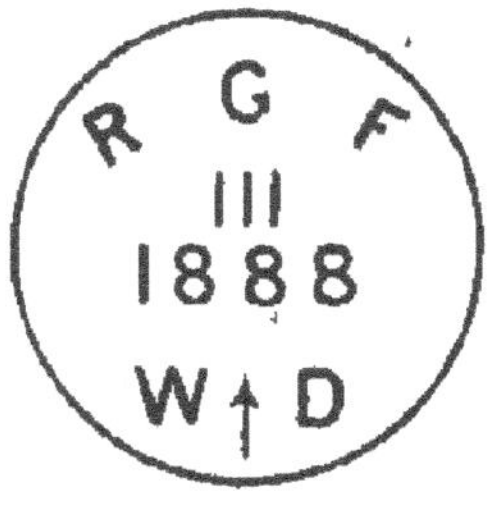

LEFT.

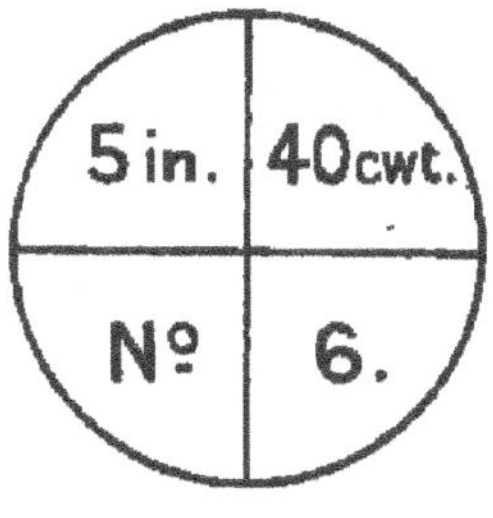

RIGHT.

The calibre mark and registered number are marked on the face of the breech, also the letter N, for naval service. A small P, surmounted by a crown, stamped on top of the gun, indicates that the gun has passed proof.

Some mention must now be made of Alpha tubes and liners, although the employment of the former has been discontinued. Alpha tubes were put in with shrinkage and employed to strengthen a gun, while liners are simply fitted in mechanically for purposes of repair when the A tube becomes worn by erosion. Shortly after the introduction of breech-loading guns of large calibre, the necessity for additional strength in the chase portion became evident. The employment of what is termed an "Alpha tube" was an alternative method of strengthening to that of the external shrinking on of chase hoops. It consists of a steel tube about half the length of the bore, rather thicker than the tubes employed for liners. Prior to shrinking on the breech-piece, the forward half of the A tube was bored out; and it was then shrunk on to the Alpha tube, the latter being further secured from forward movement by a steel collar screwed in at the muzzle.

Alpha tubes have been supplied to the 9·2-inch Mark V, 12-inch Marks V and VI, and 13·5 inch Mark II.

As the heavy breech-loading guns were introduced, the large charges employed were found to develop very rapid scoring in the barrel. To meet this it was proposed to line the chamber and first half of the bore with a thin tube made of specially well-forged steel. This "liner," as it is termed, is inserted from the chamber end, with a close mechanical fit, and is secured by a steel collar screwed into the rear end of the A tube, similarly to the way in which the "Alpha" tubes were secured at the muzzle.

The object of lining a new gun, as done at first, was to gain additional safety by protecting the A tube at the part where the highest pressures occur, and to add to the duration of the gun's first life by putting in a well-worked liner. Since 1889, however, guns are issued without liners, the dimensions of which are now considered according to the extent of wear at the end of the gun's first life, when it becomes too much eroded to properly rotate its projectiles, and is returned for repair. As the result of experiments it has been found that a gun with a cracked liner can be safely fired for many rounds with full charges, five rounds with reduced, and no less than fifty rounds with full charges having been fired from a gun after the commencement of a crack in the liner.

The description of gun construction just given refers to that which has been in general use up to quite recently, but it is being superseded by what are known as wire guns, a brief description of which must now be given.

The idea of strengthening guns by means of steel wire is not new, but to the late Mr. J. Longridge, C.E., undoubtedly here in England belongs the credit of being the first patentee, and of consistently following up and strongly advocating the wire system from its first commencement up to the present date, although he never succeeded in persuading the government to adopt his invention, and the honour of first successfully constructing wire guns belongs, like so many other of the great modern improvements in gunnery to the Elswick firm. Dr. Woodbridge, an American, claims to have originated the wire system in 1850, but the United States government have not adopted it up to the present, and there seems to be no doubt that it was Mr. Longridge who first pointed out the proper mode of winding on the wire with the initial tension so adjusted as to make the firing tension of the wire uniform for a given maximum powder pressure. In 1871, also, a Captain Schultz in France investigated the subject and designed some wire guns, several being made for the French government, but no other government except our own has as yet adopted this method of construction for guns, although several ships of the smaller Powers lately built at Elswick have been armed with wire-built quick-firing guns turned out at the ordnance works of that firm.

Between 1875 and 1879 Elswick successfully built over forty wire guns, the largest being of 10-inch calibre, but after that date the construction was stopped for some years, as the guns built in the usual manner had ample strength, while the wire system was more expensive, and, at the same time, the guns took longer to make, for with the hoop construction the hoops can all be prepared at the same time, and the different operations, at least for the smaller guns, did not take long, whereas in the early days of wire guns, the wire had to be wound slowly, and the winding could not be commenced until the barrel and interior hoops were ready. Two guns according to Mr. Longridge's designs were made, one in 1885, by Messrs. Eason and Anderson, for the English government, and the second at Aboukoff in 1888 for the Russian government, both were of 6-inch calibre, and the wire used was of riband form, one-tenth of an inch thick. Mr. Longridge advocated a thin tube, having a low modulus of elasticity, and considered that if steel wire could be obtained with an elastic limit of 80 tons on the square inch, and with a modulus of elasticity of 30,000 tons, it would be the best possible material for gun construction, when combined with a tube of modulus of about 5,500 tons (about 12,500 tons is the modulus of ordinary gun steel in the English service). At Woolwich experiments with wire guns have been carried on at intervals since 1883, but it is only since 1887 (as the result of the great success which attended the trial of the so-called "Jubilee" 9·2 in. wire gun) that it has been definitely decided to adopt the system for heavy guns.

The main difficulty in all wire constructions is the provision of longitudinal or girder strength, to resist the tendency to bend, so prominent in the long guns of the present day, the wire coil giving none, its function being solely to resist the bursting or circumferential strength. An objection has also been raised that a glancing shot might pierce the covering hoop and cut several turns of wire, thus seriously damaging the gun, whilst, with the ordinary construction, the same shot would perhaps do no more than deeply indent a hoop; but experience has shown that the wire is not so susceptible to damage as might be supposed, and it is a curious fact, that if it breaks during the winding, it unwinds very little, the friction between the parts being so great. Other objections have been that the heat imparted by firing would so alter the dimensions as to considerably modify the stresses, and also that in bending a wire over a cylinder it is impossible to give the proper degree of stretch to both its sides; but the latter disadvantage would be extremely small in the case of the thin riband form of wire now used. However, in the early part of 1890 the Ordnance Committee recommended that wire construction should be adopted, and the following were some of the reasons given:—(1) "Absolute soundness of the material employed for the reinforcement of the A tube would be obtained by the use of wire, whereas no possible system of testing, nor care in manufacture, could in all cases insure this result when forged steel hoops are used. (2) If there be a flaw, whether hidden or visible, in a tube or hoop, there is great danger of such flaw extending, until rupture takes place. Where, however, wire is used, if a rupture takes place in any coil, the adjacent coils are not affected, and this rupture cannot spread. (3) Steel in the form of wire or riband possesses double the tensile strength of steel in the form of forged tubes or hoops. (4) Such augmentation of strength, besides adding to the margin of safety under ordinary circumstances, would prove of the greatest value, if it should be desired to introduce a class of powder entailing higher chamber pressures than those for which service guns are constructed, and this augmentation in strength will render it possible to employ with safety very large charges of powder, and thus to considerably increase the power of our guns without increasing their weight."

An objection had been raised that, as to develop the full advantages of the wire system of construction some changes in the external dimensions of the guns would be necessary, great expense would be incurred in also altering the mountings, but the committee pointed out that the present guns would still continue in the service, and that, consequently, the present mountings would be utilized; but that in any case a distinct advantage in gun construction should not be lost by adhering to present pattern of mountings. As regarded the question of the construction of wire guns, the only new expense to be incurred was the providing the winding apparatus for the wire, all the old plant being still required for the manufacture of A tubes, jackets, hoops, breech-coils, etc., as the wire is merely substituted for one or more layers of hoops.

In manufacturing wire guns, the barrels, or A tubes, are manufactured in a manner similar to those for other breech-loading guns, but they are turned to a somewhat reduced diameter over the part where the wire is to be applied; a thick raised portion is left at the breech end, and shoulders or steps provided at intervals towards the muzzle, according to the design. At Woolwich the winding of the wire is performed by means of a lathe, having a side portion (carrying the wire-drum, lever, weights, and dies) so adjusted as to move sideways up and down the tube automatically as the winding proceeds; the tube (to which the wire is fastened) is turned slowly round, revolving outwards, and thus draws the wire off the drum between two dies or jaws, the upper of which is kept forced downwards by a screw-press, while the lower one is movable in the same direction as the wire, and is attached to the short arm of a lever, on the longer arm of which are weights; the wire is thus compressed tightly, and it is by the careful adjustment of the weights that the exact tension of winding on is regulated, which varies, roughly speaking, from about 35 to 50 tons on the square inch at the commencement, to about 20 to 35 at finishing off, the speed of winding being now 80 feet per minute. On the new 12-in. wire guns, the first of which were mounted in the autumn of the year 1895 in the new first-class battleships *Majestic* and *Magnificent*, which were commissioned a few months later, there are no less than 113 miles of wire; thirty-six working days (eight hours a day) being required for the winding, and from eight to nine months for the complete construction of the gun. In sectional form the wire is approximately rectangular, being $\frac{1}{4}$ in. across to ·06 in. thick, and is tested to a breaking strength on the square inch of from 90 to 110 tons. The wire on the heavy guns extends all the way out to the chase, there being fourteen layers at that part of the gun, increasing gradually to ninety-two layers over the chamber, where there is a thickness due to eight miles of wire, 8·54 miles of wire weighing a ton. The mode of fastening adopted is by means of a plug screwed tightly down on the end of the wire, which is bent cold to fit into a recess under the screw; in the same way the other end of the wire is secured after placing a clamp to prevent the tension from getting relaxed. Any interstices left in winding are carefully plugged with pieces of the wire as packing, previously filed to exact shape. The wire always commences just above the breech end of the chamber; the first layers are wound the whole length, the next lot of layers not extending so far, being fastened forward to a steel ring fitted over the layers below, and so on with the remaining thicknesses, the number of thicknesses of layers of wire being regulated throughout the length according to the gaseous pressures to be expected in that part of the gun.

CHAPTER VI.

THE DIFFERENT NATURES OF BREECH-LOADING GUNS IN THE NAVY, THEIR CHARACTERISTICS, ETC.

THERE are a large number of breech-loading guns of different calibres in use in the naval service, and sometimes as many as five or six different patterns of the same gun, the result of improvement effected in the gun since it was first designed, either in the rifling or construction, these different patterns being designated as Mark II, or Mark III, etc., as the case may be. Thus there are six different marks of the 6-inch gun, seven of the 9·2-inch, and so on. In a handbook of this kind it is not necessary, even if it were practicable, to go into details of all the various natures of naval guns. It must suffice to mention and describe briefly the latest types of guns which are mounted in the different classes of ships in the English Navy. Leaving out of account for the present the 3-pounder, 6-pounder, and 12-pounder quick-firing guns, which are intended mainly for use against torpedo-boats, there are guns of eleven different calibres, including the new 12-inch wire gun, forming the principal and secondary armaments of our ships. Their calibres are as follows: 4-inch, 4·7-inch, 5-inch, 6-inch, 8-inch, 9·2-inch, 10-inch, 12-inch, 13·5-inch, and 16·25-inch.

The 4-inch 26-cwt. gun, throwing a projectile of 25 lb., is mounted for the most part in the first-class gunboats, which carry six of these weapons. The latest pattern of this gun is ten feet long, the length of the rifling being 87·7-inches; the rifling consists of twenty-four grooves ·05 inch deep, with a twist increasing from 1 in 120 at the breech to 1 in 30 at 43·77 inches from the muzzle; the remainder uniform at 1 in 30. The gun is constructed with an A tube, over which is shrunk the jacket with trunnions, the B hoop and B tube extending to the muzzle. The projectile has a muzzle velocity of 1,900 feet per second, giving a penetration of 5·4 inches of wrought iron at 1,000 yards.

The 4·7-inch is the first type of the heavier nature of guns, constructed on the quick-firing principle, and is mounted as the principal armament of the more modern smaller cruisers, and also forms the secondary batteries of the first-class battleships *Nile* and *Trafalgar*, *Barfleur* and *Centurion*. A fuller description will be given in the chapter on quick-firing guns.

The 5-inch 40-cwt. gun is now practically obsolete. As it is only mounted in some of the old third-class cruisers, and as a drill gun for instructional purposes in naval reserve drill ships and batteries, it is useless to describe it in detail.

The 6-inch 5-ton gun is a very important weapon, having for some years formed the auxiliary batteries of many of our battleships, and the principal armament of many of our cruisers. It has been superseded by the 6-inch quick-firing gun, the latest pattern of which are wire guns, and somewhat heavier in weight, viz., 7 tons, but the projectile thrown is the same, viz., 100 lb. As the result of a series of experiments it has been found possible to alter the breech mechanism of the 6-inch guns and convert them into quick-firers; this has been a great saving of cost, and has enabled the re-armament of the ships with quick-firing guns to be effected at a far more rapid rate than could possibly have been done if entirely new guns had had to be constructed. As it is, a very large number of the 6-inch guns have now been converted into quick-firers and mounted in various vessels, although, of course, these converted guns will sooner or later be replaced by the new pattern wire gun. The latest pattern 6-inch breech-loading gun is built up of seven parts: A tube, breech-piece, 1-B and 2-B tubes extending to the muzzle, jacket with trunnions and C hoop, also a hood attached by fixing screws to the breech. The rifling consists of twenty-four grooves of the most modern rifling, known as the polygroove hook section, the twist increasing from 1 in 106·5 at the breech, to 1 in 35 at 53·125 inches from the muzzle; remainder uniform 1 in 35. The polygroove hook section is a very shallow groove to suit a soft copper band and consists of a driving side only, the curve of which is a perfect quadrant of a circle; the other side is sloped off into the bore, a small portion of which is left to form a land between two consecutive grooves. The depth varies from ·04 of an inch in field- to ·06 in the heaviest guns. The total length of the gun is 14 feet 5·5 inches, the length of the bore is 26 calibres, or 156 inches, and the length of the rifling 126·87 inches. The weight of the projectile is 100 lb., and with a charge of 48 lb. of powder it has a muzzle velocity of 1,960 feet-seconds, capable of penetrating 10·5 inches of wrought iron, at 1,000 yards.

The 6-inch quick-firing gun will be described in the chapter on quick-firing ordnance.

The 8-inch 15-ton breech-loading gun is mounted in five ships only, viz., the *Bellerophon*, one of the old broadside battleships, of which it forms the principal armament; and in the four second-class cruisers of the *Severn* class, in each of which two are mounted, one forward and one aft. It is 21 feet 2 5 inches in total length, the bore having a length of

236·9 inches (29·61 calibres), and the rifling of 195·8 inches. The rifling consists of thirty-two grooves, polygroove hook section, increasing from 1 in 110 at breech to 1 in 35 at 99·7 inches from the muzzle; remainder uniform at 1 in 35. This gun throws a projectile 210 lb. in weight, with a muzzle velocity of 2,150 foot-seconds, and is capable of penetrating a wrought iron plate 14·9 inches thick at 1,000 yards.

The 9·2-inch 22-ton gun is one of the most important guns in the Navy, as it is used as the bow and stern chaser in all our first-class cruisers. An experimental gun of this calibre was one of the first breech-loading guns manufactured in the Royal Gun Factory; it was built up of wrought iron and steel after the fashion of muzzle-loading guns, the barrel being covered with wrought iron completely from the breech to the muzzle. It was fitted with the Elswick cup obturation. This type was also the first heavy gun in which the De Bange obturation was tried, and which proved such a thorough success. The latest pattern of this gun (not the wire gun, now under construction) has been built up so as to give greatly increased longitudinal strength: above the front part of the powder chamber, the forward thrust is met by a succession of shoulders formed on the inside of the breech-piece, and similarly again in the layers above. The breech-piece and 1-C hoop (fig. 40), are secured longitudinally at the breech and by a screwed bush, to the inside of which gears the breech-screw. The A tube has also more shoulders forward, which butt against corresponding shoulders inside the B hoop and B tube; these are in turn locked back to the trunnion-ring by the 2-C hoop, so that longitudinal strength is secured throughout. The gun has four layers of metal, instead of three, as in the earlier patterns; their being consequently of reduced thickness tends to insure greater soundness of the steel, while the

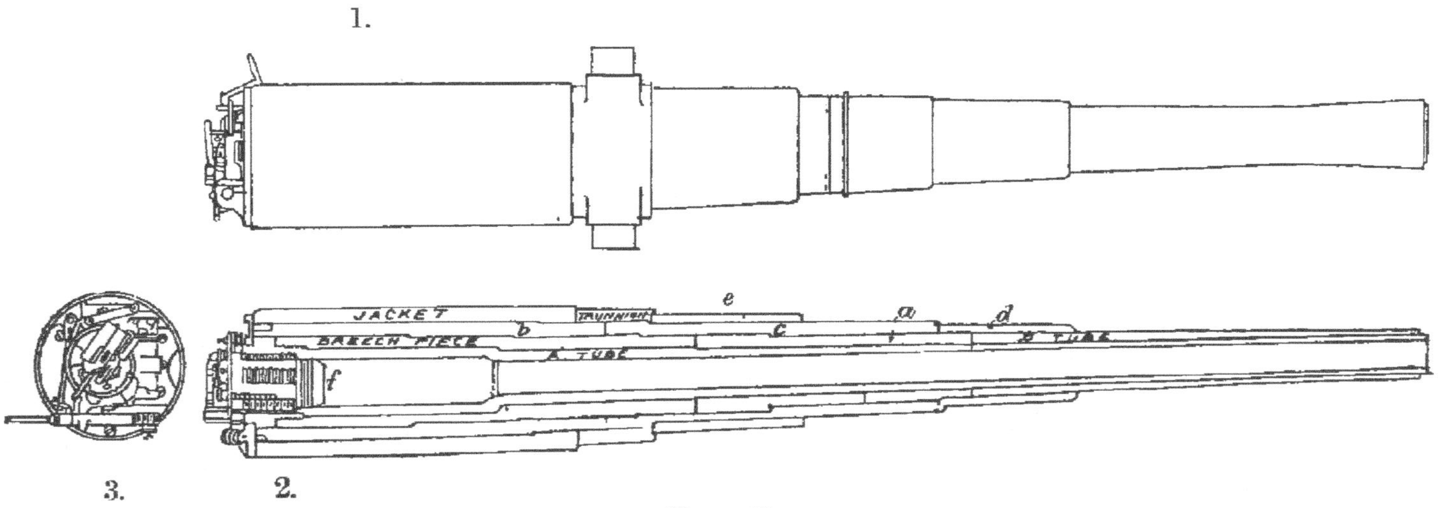

FIG. 40.

1. 9·2 inch, 22-ton gun.

2. Section, showing Method of building-up gun, with the Breech-screw and Obturator.

3. Rear View, showing Breech Mechanism.

a = B hoop ; b = 1-C hoop ; e = 2-C hoop ; d = 3-C hoop ; e = D hoop ; f = obturator.

shrinking on of a fourth layer increases the circumferential strength of the gun. The gun is built up of the following parts:—Over the A tube, in the second layer of metal, are shrunk the breech-piece, B hoop and B tube extending to the muzzle; in the third layer are the 1-, 2- and 3-C hoops, the last extending over a portion of the B tube. A thick screwed bush secures the breech-piece and 1-C hoop longitudinally at the breech end; this bush is also threaded internally for the reception of the breech-screw. In the fourth layer are the jacket, secured at the breech end by a screwed steel ring; the trunnion ring which interlocks below with the 1-C and 2-C hoops; and the D hoop, shrunk round the 2-C hoop in front of the trunnions. A bronze frame for carrying the breech mechanism is attached to the breech end by fixing screws. The gun is 25 feet 10 inches in total length, the length of the bore being 289·8 inches (31·5 calibres), and the length of the rifling 243·4 inches. The rifling consists of thirty-seven grooves, polygroove hook section, the twist increasing from 1 in 118·5 at breech to 1 in 35 at 81·12 inches from the muzzle; remainder uniform 1 in 35. The gun throws a projectile 380 lb. in weight, with a muzzle velocity of 2,036 feet-seconds, capable of penetrating an iron plate 18·3-inches thick at a distance of 1,000 yards.

The 10-inch 29-ton gun is at present only carried as the main armament in five of the battleships; the new ships *Centurion*, *Barfleur*, and *Renown*, and two of the earlier turret-ships, the *Thunderer* and *Devastation;* one of these guns is also mounted as a stern-chaser in the *Sans Pareil*. The gun consists of the following parts, viz., A tube, breech-piece, B hoop, B tube, 1-, 2-, and 3-C hoops and D hoop. Over the A tube are shrunk the breech-piece, B hoop and B tube, extending to the muzzle; over the breech-piece is shrunk the 1-C hoop, to which it is secured longitudinally by a screwed steel bush at the breech end; the bush is also prepared for the reception of the breech-screw. The 2-C and 3-C hoops are shrunk over the B hoop and the B tube. The jacket is shrunk over the 1-C hoop, and secured at the breech end by a screwed steel ring, the sight ring being shrunk round the front portion of the jacket. The trunnion-ring is shrunk over portions of the 1-C and 2-C hoops, interlocking with them; the D hoop is shrunk round the 2-C hoops in front of the trunnions; shoulders are also formed on the A tube over the powder chamber to give greater longitudinal strength. The gun is 28 feet 6·4 inches in total length, the bore 320 inches (32 calibres) and the rifling 262·18 inches. The rifling is uniformly increasing from 1 in 60 at the breech to 1 in 30 at the muzzle; there are forty grooves, ·06 inch deep and ·6 inch wide, of the polygroove hook section. The weight of the projectile is 500 lb., and the gun has a muzzle velocity of 2,040 feet-seconds, and is capable of piercing 20·5 inches of wrought iron at 1,000 yards. This is the largest gun made with trunnions.

The 12-inch 45-ton gun, which is mounted in five of the smaller battleships, viz., the *Colossus, Edinburgh, Collingwood, Hero,* and *Conqueror*, must not be confounded with the 12-inch 46-ton wire gun, which is the weapon to be mounted in the new battleships, and which will be described later on. The older gun, as originally designed, weighed only 43 tons, but on the gun trials of the *Collingwood*, in 1886, one of these guns burst, the whole chase, which was unhooped, being blown clean off; and, as a result, all that pattern were condemned, and the guns were in future hooped out to the muzzle, which although adding to their weight, immensely increased their strength. The gun is built up of the following parts, viz., alpha tube, A tube, breech piece, B tube, 1-, 2-, and 3-C hoops and 1-D and 2-D hoops. The A tube is shrunk over the alpha tube, which is secured at the muzzle by a screwed steel ring; over the A tube are shrunk, the breech-piece, which is prolonged at the rear for the reception of the screw, and the B tube. The 1-C, 2-C, and 3-C hoops are shrunk over the breech-piece and portion of the B tube, the 1-C hoop interlocking with the breech-piece and B tube. The 1-D hoop is shrunk over the 1-C and 3-C hoops interlocking with them; the 2-D hoop being shrunk over the 3-C hoop at the breech. As the gun has no trunnions, it is provided with "thrust-collars," or projecting ribs made to seat into corresponding grooves in the saddle on the mounting, thus transferring both recoil and twisting strain; the centre rib extends round only the under part of the gun, and there are two other shallow projections nearer the breech; over each of these parts is fitted a circular steel band securing the gun to the saddle; elevation and depression are given by the hydraulic gear actuating the slide, which will be explained later on. These guns are, comparatively speaking, short, their total length being only 27 feet 4·5 inches, the bore 303·0 inches (25·2 calibres), and length of rifling 250·8 inches. There are forty-eight grooves in the rifling, with twist increasing from 1 in 120 at breech to 1 in 35 at 126·27 inches from the muzzle, remainder uniform 1 in 35. The gun fires a projectile 714 lb. in weight, and with a muzzle velocity of 1,914 feet-seconds can pierce 20·4 inches of wrought iron at 1,000 yards.

The 13·5-inch 67-ton gun is an important weapon, forming as it does the principal armament of many of our first-class battleships; four of these guns are mounted either in turrets or barbettes in the *Royal Sovereign*, *Empress of India*, *Repulse*, *Resolution*, *Ramillies*, *Revenge*, *Royal Oak*, *Hood*, *Nile*, *Trafalgar*, *Anson*, *Camperdown*, *Rodney*, and *Howe*.

The gun is made up of an A tube, breech-piece, B hoop, B tube, 1- and 2-C hoop, C tube, 1- and 2-D hoops and jacket. Over the A tube are shrunk the breech-piece, B hoop and B tube, extending to the muzzle. Over the breech-piece is shrunk the 1-C hoop, which are secured longitudinally to each other and to the A tube by a screwed steel bush at the breech end; the bush is also prepared for the reception of the breech-screw. The C tube and 2-C hoop are shrunk over the B hoop and B tube. The 1-D hoop is shrunk round the 1-C hoop and a portion of the C tube, and is secured at the breech end by a screwed steel ring. The 2-D hoop is shrunk over the C tube immediately in front of the 1-D hoop. The jacket is shrunk round the 1-D hoop at the breech. Thrust collars are formed on the exterior of the 1-D hoop for securing the gun on the mounting. The total length of the gun is 36 feet 1 inch; the length of the bore 405 inches (30 calibres), and of the rifling 333·4 inches. There are fifty-four grooves, the rifling having an increasing twist from 1 in 120 at the breech to 1 in 30 at 166·7 inches from the muzzle; remainder uniform 1 in 30. The gun fires a 1,250 lb. projectile, with a charge of 630 lb. of powder giving a muzzle velocity of 2,016 feet-seconds, and a penetration at 1,000 yards of 28·2 inches of wrought iron. It is doubtful if the manufacture of these guns will be continued, the new 12-inch 46-ton wire gun will in all probability be substituted eventually for them.

The 16·25-inch 111-ton gun is the largest gun which has been constructed in this country, and it is unlikely that any more guns of anything like this size will ever be constructed again, at least for naval service. Two of these guns have been mounted as the main armament in three ships only, the *Benbow*, *Sans Pareil*, and the unfortunate *Victoria*, which was sunk by collision with the *Camperdown* off the coast of Syria on the 22nd June, 1893; in the *Sans Pareil*, as also was the case in the *Victoria*, the two guns are in a single turret forward; in the *Benbow*, they are mounted singly in barbettes, one forward and one aft. These guns, of which some are also to be found in the Italian Navy[1] were designed and made at Elswick. The gun is made entirely of steel (in five layers) and consists of an A tube, around which are shrunk the breech-piece and seventeen hoops extending from the breech to the muzzle, the breech-piece being prolonged at the rear for the reception of the breech-screw, and secured to the A tube longitudinally by a serrated ring of metal. Over the breech-piece and the adjoining hoops are shrunk the third, fourth, and fifth layers, a series of twenty-six hoops, which are secured longitudinally by a serrated ring of metal. The forward hoop of the fifth layer forms the "thrust-ring," and, similarly to the breech-piece, this is held longitudinally to the layer beneath by a serrated ring of metal; on the exterior of the thrust-ring "collars" are formed which fit into corresponding grooves in the cradle. Two shallow bands are also formed on the exterior nearer the breech end of the gun. The end of the breech is rounded off and the muzzle has a more decided swell than in the case of the 13·5-inch guns (fig. 41). After proof and subsequent firing trials some of these guns were found to droop slightly at the muzzle, and bend very slightly to the right, also some of the forward hoops had moved a little apart, so certain

[1] Four of these guns are mounted *en barbette* at the fore and after ends of a central armoured casement in the first-class battleships, *Andrea Doria*, *Francesco Morosini*, and *Ruggiero di Lauria*. These guns are slightly different to the English 111-ton gun, as they only weigh 105 tons, while the calibre is 17 inches instead of 16·5 inches, and the weight of projectile 2,000 lb. instead of 1,800.

modifications and additions in the shape of longer hoops and putting on additional long hoops have had to be made. The gun weighs 110 tons 12½ cwt., and its total length is 43 feet 8 inches; the bore is 487·5 inches long (30 calibres), and the rifling is 397·2 inches; there are seventy-eight grooves of the polygroove E. O. C. section ·06-inch deep and ·45 in width; in the latest pattern of these guns the rifling increases throughout from 1 in 60 to 1 in 30. The chamber has a diameter of 21⅛ inches, length 84½, and capacity 28,660 cubic inches. The weight of the projectile is 1,800 lb., and with a full charge of 960 lb., giving a muzzle

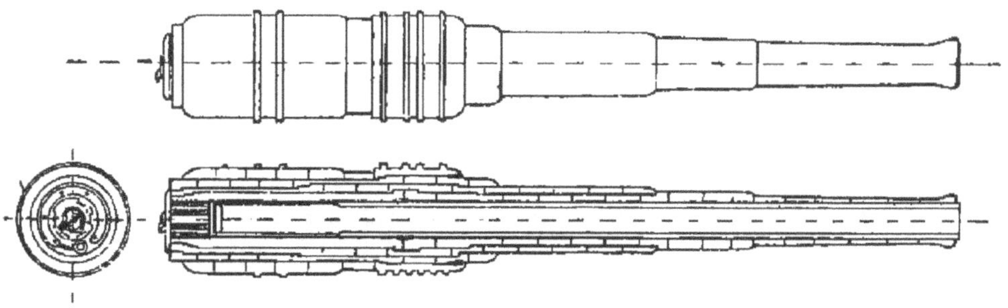

FIG. 41.—THE 16·5-INCH, 111-TON GUN.

velocity of 2,087 feet-seconds, it will penetrate at 1,000 yards range 32 inches of wrought iron.

As it may be asked why the authorities have decided to discontinue the manufacture of these very heavy guns, and to arm our new battleships with weapons of only a medium size, it may be as well to point out briefly one or two of the objections which militate against the use of these enormous weapons for the armament of ships. For one thing it has not been found possible to mount more than two of these 111-ton guns in a ship, except at a great sacrifice of efficiency and protection in other directions. It is true that the Italians have three battleships, each of which carries four 105-ton guns, but they have also come to the

West & Co., Southsea, photo.

To face p. 122.

H.M.S. "BENBOW" SHOWING 111-TON GUN IN AFTER BARBETTE.

conclusion that these guns are a mistake, and all their new ships are to be armed with a medium-sized gun; the French similarly have given up the 75-ton guns mounted in their older first-class battleships and have substituted a 44-ton gun for their new ships. If we compare the *Benbow* and the *Anson*, two practically sister-ships except in their heavy armament, the latter carrying four 67-ton guns instead of the two 111-ton guns of the former, it will be seen that the *Benbow*, although her guns throw a projectile 550 lb. heavier than the *Anson's* guns, is yet at a disadvantage in the total weight of metal she can discharge at an enemy, for her two guns together only throw 3,600 lb. against the 5,000 lb. thrown by the four guns of the *Anson*. Then again, the 67-ton guns of the *Anson* can be worked quicker than the 111-ton guns of the *Benbow*, for under favourable circumstances two rounds can be fired in a couple of minutes or even less from the 67-ton guns, while it takes two minutes and a half for two rounds from the 111-ton guns. And this increased rapidity of fire with four projectiles against two would in all human probability be a factor of considerable importance in favour of the ship carrying the four guns as against the vessel with only two, notwithstanding their increased size. A third point of much importance where these heavy guns are concerned, lies in the fact that the bigger the gun the shorter is its life, and the life of the 110-ton gun cannot be reckoned at more than from seventy-five to eighty full charges (what used to be called battering charges are now termed full), *i.e.*, if an average gun of that size has fired seventy-five full charges, its projectile might not be rotated, and would consequently be liable to a complete loss of accuracy; this result being due to the bore at the end of the chamber gradually wearing, so that after a certain number of rounds—varying for each class of gun and each individual gun—the copper band on the projectile overrides the rifling. The wear and enlargement may not extend far into the bore, but the projectile will soon acquire considerable velocity, so that even when it passes the worn part, the object of the rifling is defeated, for the driving band will not have strength enough to alter the motion the projectile has acquired and will "strip" rather than impart rotation to it. To partially meet this difficulty, or rather to postpone its consequences, long strips of copper, called "augmenting strips" are supplied, to be hammered partially into the cannelures on the driving bands, and thus augment their diameter. When further wear has rendered this remedy inoperative, the gun must go back to the factory to be re-lined. The larger the charge of powder used, the sooner will the gun wear, for it is the actual amount of gas that erodes and wears the barrels. Moreover, some powders have a much more erosive action than others, this action depending a great deal on the heat developed; both these conditions are particularly unfavourable to very heavy guns, which necessarily require large charges, and the powder at present used with them gives out more heat during combustion than any other. With the 67-ton gun from 100 to 110 full charges could probably be fired before it would be necessary to use augmenting strips with the projectiles, while from a 6-inch gun from 400 to 450 rounds can be fired. It has to be remembered, however, that arrangements are made for using reduced charges, and it is only in special cases that the full charge is used. As has been already briefly explained, the usual method of repair is to bore out the barrel and then insert a tube or liner. The liner may extend the whole length of the bore or for only a portion of it. After it has been inserted, the liner is expanded, to accurately fit the bore, by firing a charge in it; it is then fine bored and rifled, and if the operations are successfully carried out, the gun ought to be practically as good as new. The comparatively speaking short life of modern guns would seem, however, to render it imperative that a large reserve should be created in readiness to meet the exigencies of war.

In the previous chapter a short description of the new wire guns was given, and a few further details of those already in use in the Navy will be of interest. Wire is now used for the construction of the 4-inch, 4·7-inch, and 6-inch quick-firing guns, also for the new 9·2-inch 25-ton, and the 12-inch 46-ton guns; the latter gun is the weapon which has been definitely adopted by the Admiralty for the heavy armament of the new battleships of the *Majestic* and *Canopus* classes, and it is built up as follows :—A tube and A outer tube, which is not shrunk, but driven on; next comes the wire, of which there is between 110 and 113 miles, wound on over the whole length of the gun in layers varying from fourteen over the chase to ninety-two over the powder chamber; then the B tube over the chase, the jacket and the C hoop, which is screwed on to the jacket and over the end of the B tube, and is practically a nut. The jacket and outer parts are put on without any appreciable shrinkage as regards the wire coil, but in order to get as close a mechanical fit as possible, slightly heating is usually necessary. The total length of the gun is 37 feet 3 inches, length of the bore, including chamber, 35·4 calibres. The rifling is straight for 84 6 inches, and then for 278·9 inches an increasing twist from 0 to 1 in 30 calibres, and there are forty-eight grooves. The full charge is 167 lb. eight oz. of cordite, and the weight of the projectile 850 lb., which, with a muzzle velocity of 2,400 feet-seconds is capable of penetrating 34·6 inches of wrought iron at 1,000 yards, and 31·2 inches at 2,000 yards. It at present takes from eight to nine months to make one of these guns.

The new wire 9·2-inch gun weighs 25 tons and has an extreme length of 32 feet; it consists of the inner A tube, outer A tube; then wire for about half the length of the gun in layers varying from 10 to 65; next the B hoop and the B tube, which extends from the muzzle to two feet over the wire, and the jacket locked to the B tube by the trunnion ring. The gun throws a projectile 380 lb. in weight, and with a cordite charge of 63 lb. has a muzzle velocity of 2,380 feet-seconds, capable of penetrating 24·5 inches of wrought iron at 1,000 yards, and 19·4 inches at 2,000 yards. The Woolwich authorities have proposed the construction of a 9·2-inch gun, 40 calibres in length, to weigh 27 tons, which will fire a cordite charge of 94 lb., giving a muzzle velocity of 2,700 feet-seconds, but the plans have not yet been approved. The first of the wire 9·2-inch guns have been mounted in the new first-class cruisers *Powerful* and *Terrible*.

Mention has been made in a previous chapter of the system adopted for closing the breech in English guns, but some further details, with the accompanying plans, may help to make the method plainer to the ordinary reader, who has had personally no opportunity of seeing the breech-action actually working. The system adopted, as has been already stated (pp. 84, 85), is that known as the interrupted screw; the breech-block being of solid steel furnished with a screw-thread of the requisite strength and pitch, while the gun is prepared with a similar female screw to receive it. In the 6-inch and smaller guns, for instance, the breech screw has three portions of the thread removed longitudinally, each one-sixth of the circumference; the interior of the gun being similarly prepared, admits of the screw—when the raised portions are placed opposite the smooth surface of the gun—being pushed home and locked by the sixth of a turn. The screw has hinged to it a cam lever (fig. 42), by which it is locked and unlocked; encircling the rear end of the breech-screw, and hinged to the hood, is a carrier-ring, which supports the breech-screw when withdrawn; the cam portion of the lever (when the breech-screw is locked) falls into a recess in the carrier-ring, and so prevents any movement of the breech-screw during firing. In depressing the cam lever after the breech-screw is unlocked, the cam acting upon the surface of the carrier-ring gives the requisite power for starting the withdrawal of the breech-screw together with the obturator, some

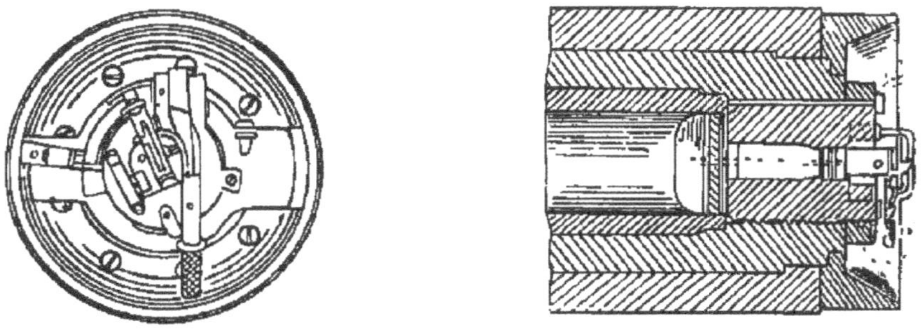

FIG. 42.—6-INCH HAND-WORKED BREECH MECHANISM. WOOLWICH DESIGN.

power being often required, as the De Bange pad has an inclination to stick after the gun is fired. The carrier-ring is held to the gun during the withdrawal of the breech-screw by a clip, and a stop-bolt prevents the breech-screw being disengaged from the carrier when withdrawn; the whole being then swung clear of the breech opening to admit of loading, the carrier-ring being retained in the loading position by a spring latch. The gun cannot be fired before the breech-screw is properly home and in its locked position. Although in these smaller guns the cam lever is made of sufficient length to give the power required for opening the breech easily and expeditiously, yet in the larger nature of guns, the 9 2-inch and onwards, something more is required. The details of the mechanism for working the breech-screws of the larger guns vary a good deal; but there are two broad classes of breech arrangements, viz., those worked by hand-gear and those worked by hydraulic or electric power. With breech-screws weighing several hundredweight, machinery of some sort had to be used if the guns were to be worked with rapidity—a matter of extreme importance; for the higher the speed of firing, the greater the power and efficiency of a ship. Hydraulic breech mechanism was originally applied to the 12-inch 45-ton guns and upwards, but such advance has now been made that the hand-worked system has been extended to the new 12-inch wire 46-ton gun, and one man can open and close the breech without difficulty in a few seconds; the work of unlocking, unscrewing, withdrawing and swinging the breech-screw clear being all performed by one lever; the mechanism being actuated by a small wheel with a winch handle in the 12-inch guns, and in the 9·2-inch wire guns, to which similar mechanism has been fitted, by a crank handle with counterweight. Similarly the hand mechanism for the old pattern 9 2-inch and 10-inch guns has been extremely simplified, the number of motions having been reduced from four (as in the older types and smaller guns) of 6-inch to three in the larger guns. To open the breech the cam lever is lifted, as with smaller guns, but the act of lifting it connects it to a very powerful form of ratchet lever, known as the "Stanhope" lever; with this appliance the breech is unscrewed, after which the operator has only to turn the winch or lever handle, A, on the left of the gun. And this, revolving the hinge bolt and pinions, causes the withdrawal of the breech-screw by means of the rack on its side; when the screw is just clear of the breech the end of the rack is reached, and at this moment the screw becomes firmly attached to the carrier by means of

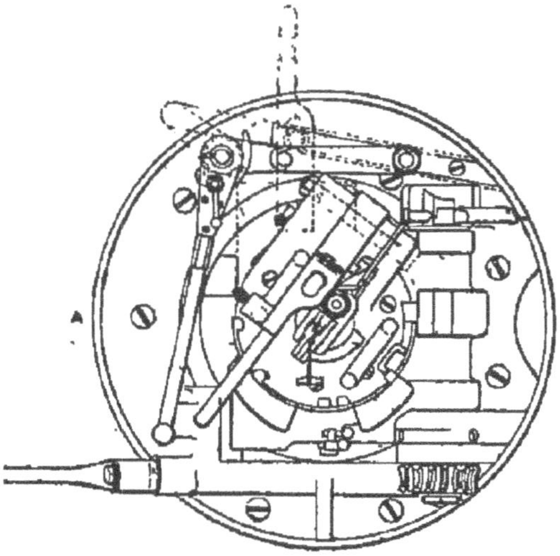

1.

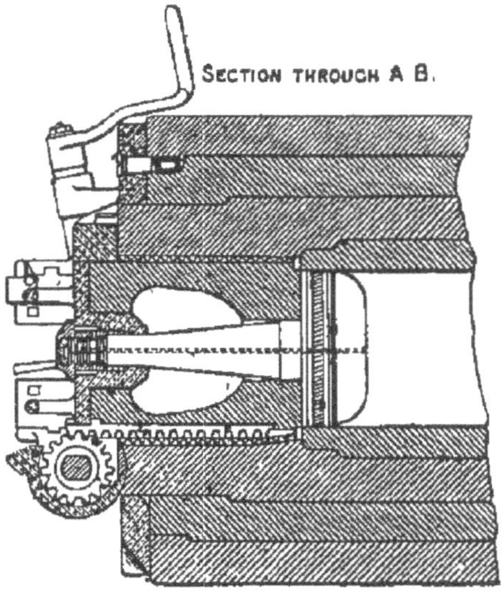

2.

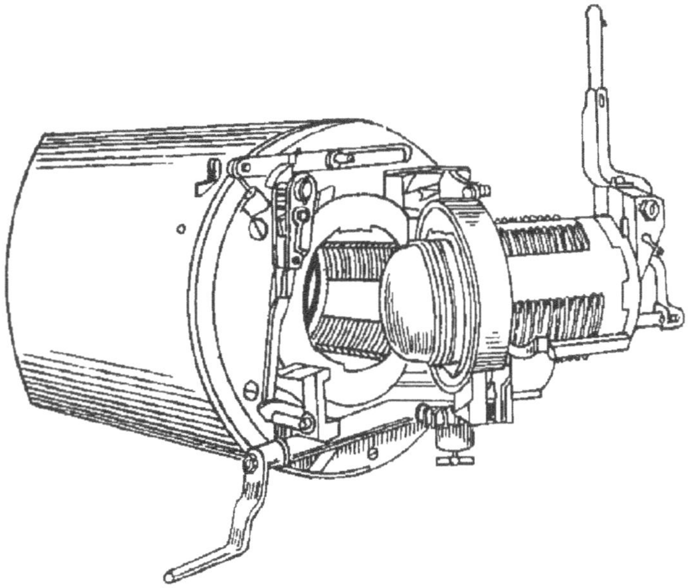

3.

FIG. 43.—9·2-INCH AND 10-INCH HAND-WORKED BREECH MECHANISM. WOOLWICH DESIGN.

1. Breech closed. 2. Breech closed, section through A B.
3. Breech open, screw swung half clear.

a stop-bolt on the right, and on the left by a clip; then the effect of continuing to revolve the pinion is to turn the whole, screw and carrier together, and swing them round to the right into the loading position (fig. 43). After completion of the loading, turning the lever handle round in the opposite direction reverses the process, and immediately the carrier comes against the face of the breech the clip is withdrawn from its recess in the screw, which is thereby freed from the carrier and is by its rack pushed home into the gun.

In the heavier guns, where the breech-screw is worked by hydraulic power, the mechanism is somewhat complicated, and only a general idea of the system of working can be given; the following description being mainly drawn from Commander Lloyd's and Mr. Hadcock's valuable work on modern artillery, to which reference has already been made. The breech-screws are provided with De Bange pads, and are on the interrupted screw principle, like the smaller hand-worked guns. As the mechanism for working the screw is fixed, it is necessary after each round to give the gun extreme elevation, and to run it in against the rear buffers; these buffers are required to have some elasticity in them, to provide for the gun striking them heavily, in which case they may to some extent absorb the blow in yielding; but if the gun is not placed very precisely each time that it is brought into the loading position, the mechanism will not gear properly into the breech-screw, and there will be trouble in working it. The difficulty is fairly well met by constructing the buffer so that it will always compress a small amount, after which further compression is prevented by a rigid pin passing through it. When in the loading position the gun rests against this pin.

The breech mechanism is worked by two levers, which must always be moved in the same direction as the machinery they control. An ingenious system of interlocking gear, proposed by Lieutenant (now Commander) Peirse, of H.M.S. *Excellent*, renders it impossible to move one lever unless the other occupies the correct position. As soon as the gun is placed in the loading position, a claw connected to the traversing portion of the breech mechanism is brought from left to right; it engages with a projection on the rear face of the breech-screw and unscrews the breech. The turning of the screw brings the projection just mentioned into gear with an extractor. This completes the first motion. The other lever is now pushed in the rear direction, and the extractor is caused to move rearwards, taking the breech-screw with it.

The third motion (worked by the lever first moved) traverses the breech-screw over to the left, out of the way.

The fourth motion pushes the loading tray into the gun, by means of the same machinery that drew the breech-screw out (figs. 44-45). Loading then takes place; after which the loading tray is withdrawn, the breech-screw traversed, pushed in, and screwed up.

The whole of these operations are done by two cylinders, known respectively as the "traversing" and the "withdrawing" cylinders. The former is generally fixed, and has a moving piston; the latter generally moves, and has a fixed piston.

In the accompanying plates, A represents the breech-screw, and E the withdrawing shaft. The drawing of the mechanism in elevation shows the portion which moves from the right to the left, or *vice versâ*, in performing the operations of locking or unlocking the screw, etc. The rear view illustrates the movements of the portions of the loading tray. It will be observed the "mechanisms" are right and left, the reason for this being that there is only

FIG. 44.—HYDRAULIC BREECH MECHANISM FOR 67-TON GUNS. END VIEW.

Breech-screw withdrawn. Tray placed ready for loading.

A A = Breech-screw. B B B = Loading tray. D = Withdrawing cylinder. E = Withdrawing shaft.

room on the centre line of the turret for the stroke of the withdrawing cylinders.

Directly the breech is opened, the chamber of the gun must be washed out, to insure the extinction of any hot residue from the last charge before the introduction of the new one. The washing out is rapidly and conveniently

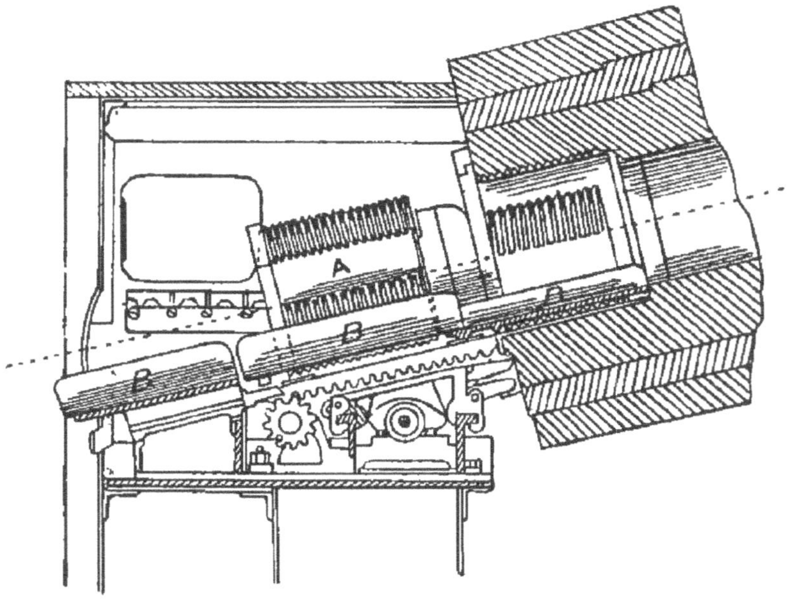

FIG. 45.—HYDRAULIC BREECH MECHANISM FOR 67-TON GUNS, SECTION.

Breech-screw withdrawn. Tray placed ready for loading.
A = Breech-screw. B B B = Loading tray.

done by a small pump with hose and nozzle, provided specially for the purpose.

It has already been mentioned that with fixed loading arrangement the gun must be brought to an exact position of training before the loading can be performed. It is also necessary that there should be no chance of the turret being moved until the loading is complete. The rammers are attached to the ship, and if the turret or turn-table in a barbette were turned while the rammer was extended for loading, it is easy to see what the consequences would be. Arrangements are therefore made for locking the turret or turn-table as soon as it is trained into the loading position, and this is done by means of two bolts. One, called the "inside" bolt, assists to bring the turret or turn-table to the proper position, while the other, called the "outside" bolt, locks and holds it during loading. Both bolts are, in all the latest designs, worked by hydraulic power. The inside bolt is a heavy bolt with a tapered end, which is lowered, by the man attending it, as the turret or turn-table approaches the loading position, until its end is rubbing on a steel plate secured to the deck; when the turret or turn-table reaches the loading position, the bolt at the same time reaches and drops into a hole with sides made to correspond with its tapered end. If the turret or turn-table is moving too fast, the bolt, owing to its shape, will ride up again, but it will have caused a check sufficient to inform the man in charge that he has passed the loading position. He would therefore reverse the training engine, and turn the turret or turn-table slowly back, until the bolt enters again; as soon as this is done, but not before, the outside bolt, which is cylindrical, can be dropped into the hole provided for it. Indicators show the exact position of the turret, and the man attending the "inside" bolt has only to watch until the pointer comes opposite to a mark, to know when to lower his bolt. Similarly, when the "inside" bolt is home, another indicator informs the man outside the turret that he can insert the "outside" bolt; then, when this bolt is home, another indicator marks it, and the loading can commence. The breech can be opened and the chamber washed out as the turret or turn-table is being trained and locked.

CHAPTER VII.

NAVAL GUN-MOUNTINGS OF THE PRESENT DAY.[1]

GREAT as has been the advance in guns made during the last thirty years, still more wonderful is the revolution which engineering skill and science has brought about in the methods of mounting them; for while the powerful rifled ordnance of to-day have been gradually evolved from the old 32 and 68-pounders, there is absolutely no connection between the old wooden truck carriages, crude relics of the Middle Ages which prevailed up to even a few years ago, and the latest pattern gun-mountings, which are machines of extraordinary ingenuity, designed to deal with enormous forces in the most scientific manner.

Rapidity of fire is almost as dependent on the mounting as on the gun itself, and convenient and rapid means for training and elevating are absolutely necessary if the accuracy of a gun is to be of service.

In the manufacture of mountings, as in that of guns, cast and wrought iron have entirely disappeared. Gun-metal, manganese bronze, or some other alloy of copper, and cast and forged steel, are the only materials now employed. Gun-metal is a mixture of tin, zinc, and copper, in about the proportion of tin, 10·8, zinc, 2·7, and copper, 86·5.

[1] The description of gun-mountings is largely compiled from Commander Lloyd's and Mr. Hadcock's "Artillery, its Progress and Present Position."

It is especially suitable for casting purposes, and for this reason is employed for intricate fittings; but its cost is high as compared with that of cast steel, and this limits its use. It is also employed for bearings of steel shafts or wherever there are steel working faces, and its anti-corrosive properties are of great advantage, especially for fittings that are not very accessible.

Manganese bronze is an alloy of copper that has lately come much into use. It apparently has all the advantages of gun-metal, together with considerably greater strength. It is, however, rather more expensive, but where a hard anti-corrosive surface is required, it is specially suitable. There is more trouble in casting manganese bronze than gun-metal, but, if successful, the castings are as sound and more homogeneous than with the latter metal.

The process of steel casting and of steel manufacture has made great progress, and is further rapidly developing, but for anything of an intricate nature connected with gun-mountings cast-steel is still avoided; not on account of its unsuitability for the work it will have to perform, but because the occurrence of blow-holes cannot as yet be prevented, and these are often only discovered after much time and labour have been expended on the machining of a casting, but may yet necessitate its rejection and having all the work to do over again. Where the design is a simple one, as, for instance, with lower roller paths, there is not much fear of failure, and large numbers of castings are passed as sound and good for every one that turns out bad; but to withstand hard blows or sudden concussions, forged material is much more reliable than cast, and therefore gun-mountings are made as far as possible of forged or rolled material.

One of the most important duties which a modern gun-mounting has to perform is absorbing the energy of the recoiling gun; and here it is that one of the greatest advances has been made, thanks to the skilful application of the science of hydraulics, for from the 3-pounder to the 111-ton gun it is the hydraulic system which does this important duty, and so successfully, that it seems unlikely that any better system will supersede it, as it unites the highest efficiency with the greatest simplicity.

It was at Woolwich that the first recoil press or buffer was applied to a gun, in the form of a hydraulic buffer which should offer a certain amount of resistance to recoil, and act as an auxiliary to the frictional compressors then in use. The arrangement consisted of a cylinder fitted with a piston and piston-rod. In the piston four holes were made; if the cylinder were charged with oil, or other liquid, and the piston forced from one end to the other, the liquid could only pass the piston by means of the holes. The velocity of the liquid was determined by the speed of the piston, and the total area of the holes in it. As the latter remained constant, it is evident that the velocity of the liquid varied directly with that of the recoiling gun. The pressure, therefore, which caused the velocity, varied continually during recoil, and when at its maximum reached a high point. Various improvements, by which it was sought to keep the velocity of the liquid the same throughout the recoil, and thus to obtain a uniform pressure in the recoil press, were designed. It was, however, a plan by Mr. Vavasseur (of Sir W. Armstrong and Co.), which, after competitive trial, proved to be the most suitable for ordinary hand-worked mountings, and it was accordingly adopted for the English service and nearly all over the world. By this scheme the piston was fitted with a valve which gradually closed the orifice or port, the movement of the valve being controlled by studs fitting into rifled grooves in the cylinder. The piston in the recoil press consists of two parts; the fixed piston made in one solid piece with the steel piston rod, and the valve made of gun-metal. In the earliest designs the peculiar shape of the port (required to obtain a constant velocity of liquid) was cut in the steel piston, and the orifice in the gun-metal valve was made plain, but in later designs these arrangements have been reversed. With the exception of the orifice and the studs on its periphery, the valve is a plain disc with a hole in its centre. It is passed on to the piston rod and kept in place by a nut, or the piston is divided into two parts, and the valve is put on in halves between them. The rifling in the cylinder gives the valve a rotation of 60° during recoil, so that when the gun is in the firing position the port in the piston is open, being opposite the orifice in the valve, but it is gradually closed during recoil by the rotation of the valve. There are regular formulæ for calculating the opening in the port at any given moment, by which its actual dimensions and shape are obtained, and also for the areas of the pistons. The Vavasseur valves are set in the following manner:—the valve is first secured to the piston, in the shut position, by a stud; the mounting is then run back as far as it can go (the valve turning the piston and piston-rod during the movement), and is then run out 2 inches. In this position the piston-rod is marked for the key which is to fix it to the bracket on the slide, but in the British service it is an invariable rule not to cut the key-ways in piston-rods until after the gun has been fired in the mounting.

The Vavasseur valve possesses the advantage of being capable of adjustment if any changes in the ammunition, or in the conditions under which the mounting is to be used, are made. A slight turn of the piston-rod will either shorten or lengthen the recoil according to the direction in which that turn is made. It was the custom, at first, to mark the piston-rods for different lengths of recoil, but it has since been found best not to invite an alteration which, if not carried out properly, would probably cause disaster. A small disadvantage in the valve is, that it occupies a good deal of space in the recoil-cylinder. In ordinary mountings this is no difficulty, but in the quick-firing mountings, where space has to be carefully studied, the Vavasseur rotating valve has been abandoned and the key-valve adopted instead, which consists of a port in the piston, which travels during recoil along a key or bar shaped so as gradually to close the port.

A question of considerable importance is the best oil to use in the presses or buffers, as several conditions have to be considered. It must be of a non-freezing quality, and have no injurious effect on the leathers; it must be sufficiently viscous, free from acid, and its specific gravity should not be high; further, highly inflammable oils are too dangerous to be used on board ships. It has been found that Rangoon oil best fulfils these conditions. It has therefore been designated "buffer oil," and the valves are calculated for this liquid. Where Rangoon oil cannot be obtained, and the temperature is warm and equable, olive and vegetable oils are also quite suitable, but they are much affected by variations of temperature, and easily frozen. Water must never be used without soap, oil, or some similar lubricant. Naval officers have a great objection to the use of oil in recoil presses, as it is impossible to keep the glands so tight but that drops occasionally escape, and oil most sadly stains the decks; but still, in these days, efficiency has to be considered before appearance, and there are objections to the use of water and soap, the only alternative which will not stain wooden decks; but if the recoil presses are filled with water and soap, the greatest care must be taken not to let the mixture freeze.

There must always be two systems of applying hydraulic buffers. Either recoil must force a piston and piston-rod further into, or it must tend to withdraw them from the cylinder. The systems are generally known as the "push in" or "compression," and the "pull out" or "tension" buffers respectively. With the first, it is evident that the cylinder must not be completely filled with liquid before recoil takes place, but that room must be left for the entrance of the piston-rod. The original Woolwich buffers were often of this pattern, and several were burst on account of too much liquid being put into them. The "pull-out" presses are free from this danger, but in them the available area of piston against which the pressure of the liquid acts during recoil is diminished by the sectional area of the piston-rod. A "pull-out" buffer requires either a higher pressure or a recoil press of larger diameter than a "push-in" one. Six-inch and larger mountings constructed on Mr. Vavasseur's original design combine the two systems, having a "push-in" press on the left, and a "pull-out" on the right. The volume of the two presses taken together remains constant whatever the position of the gun, for as one piston-rod enters, the other is withdrawn. The cylinders are joined together by a pipe, and the liquid displaced on the left side by the entering rod passes through this pipe into the cylinder on the right. In order to be able, if required, to hold the gun "in" at extreme recoil, a non-return valve is placed in the connecting pipe, so that, although the liquid can pass from left to right, it cannot leave the right-hand cylinder by the same passage, and the piston-rod, which is withdrawn during recoil cannot re-enter its cylinders. A valve called the "bye-pass valve" supplies the means of letting the gun run out when desired, and it opens or closes a passage for the liquid independent of the non-return valve, and is worked by hand only. Under ordinary circumstances the "bye-pass" is left open, and the gun, therefore, runs out automatically after recoil; but if it is required to keep the gun in, it is closed before firing. The Vavasseur hydraulic recoil press fulfils all the conditions required to control a gun. Perfectly automatic in action, it brings no blow on the mounting; it gives complete control over the gun for running in or running out; and although with breech-loading guns it is no longer a matter of great importance, it regulates the length of recoil with the greatest possible nicety, no matter what the charge is, for the recoil always continues until the valve is closed. To Mr. Vavasseur are due also other very important improvements, which were introduced at the same time as his recoil presses. Breech-loading guns having been adopted, it became necessary to provide for loading the guns out in the firing position. In Mr. Vavassenr's carriages, the system of making the guns recoil up an inclination is retained, so that after the guns are brought to rest they run out again by gravity, and thus automatically acquire the "out position." The recoiling carriage consists simply of the two recoil cylinders, which are placed on each side of the gun, and joined together by a front transom. The slides, on which the carriage rests, are raised so as to carry the gun at a convenient height, and to give room for the breech when the gun has extreme elevation. The rope-breechings, which had been in existence from the earliest days of gunnery, were finally done away with when the Vavasseur mountings were introduced.

The system of elevating in the old mountings was always a source of delay and trouble. As the elevating gear was all attached to the recoiling carriage, it was unsafe to handle it when the gun was "ready" for firing. If, therefore, the elevation had to be altered, the tube or primer had to be removed from the vent before the required correction could be made, and then the gun had to be made "ready" again before it could be fired; while all this took so long, that it frequently happened that before the captain of the gun could fire, a further alteration in the trim of the ship, necessitated another readjustment of the elevation. A system by which the elevation could be corrected up to the moment before firing, was, therefore, a most welcome improvement; and this is one of the special features of the Vavasseur mountings. The shaft of the elevating hand-wheel is attached to the slide or non-recoiling portion of the mounting, and is fitted with a long key-way. A sleeve or tube carrying a worm is placed on the shaft, so that, although a key compels the sleeve and worm to "turn" with the shaft, they are free to slide along it when the gun recoils, the shaft being exactly parallel to the motion of recoil. The worm drives the elevating-gear, and is held in position by brackets formed on the carriage. The training arrangement in the Vavasseur system gives complete control over the mountings by means of worm gear; and the rope tackles, used for so many centuries for working and controlling gun-carriages, have been entirely superseded.

In the first made Vavasseur mountings, which were designed for use in gun-ports on the broadsides of ships, and which are designated officially V. B. (Vavasseur broadside), the cast steel slides are provided with a pivot bar connecting them to the pivot in the centre of the port. Rollers attached to the base of the slides run on racers let into the deck, as with previous mountings. A stout clip plate is fitted to the front of the slide, and, in connection with a clip ring, secured to the deck, provides for the lift. This is necessary on account of the leverage at which the force of recoil acts about the pivot. The clip ring is also con-

6-INCH GUN ON VAVASSEUR BROADSIDE MOUNTING.

structed so as to take the recoil, and it is thus perfectly safe to fire the guns without the pivot pin or bar. The tops of the slides are inclined to a considerable angle, ranging from 15° for the 4-inch and 5-inch, to $7\frac{1}{2}$° for the 6-inch broadside mounting. Rollers are fitted to all slides inclined at less angles than 10°, so as to reduce friction and insure the automatic running out of the top carriage. The 6-inch mountings were designed to "house" or run in so far that the muzzle of the gun should be within the ship. The slides accordingly provide for a recoil of 42 inches. The bye-pass will hold the gun in after recoil, but running in can also be performed by means of a long screwed shaft on the left of the mounting. A connection can be made between the top carriage and this shaft, so that turning the latter screws the carriage and gun to the rear, but care must be taken not to fire the gun with the connection to the shaft in gear. In these broadside mountings, as at first designed, the top carriages consisted of the two cast steel buffers or cylinders, and of a transom and bottom plate joining them together. The recoil valves had two ports and two studs.

The "centre-pivot" mountings (fig. 46), known as the V. C. P. (Vavasseur centre pivot), were at first introduced, a little later than the broadside, for special positions; but the immense advantages they possess have caused them almost entirely to supersede the "broadside" mountings. The only work to be done in training, whatever the inclination of the ship, will be due to friction. In centre-pivot mountings the pivot, racers, training gear, and clip ring can all be made self-contained, which is in itself a great convenience. Alteration to the form of the deck often gives considerable trouble with broadside mountings, a trouble that is never experienced with centre-pivot mountings; although if a small gun-port is required in the side of the ship, the broadside or front-pivot mounting becomes a necessity. Sponsons projecting from the sides are now often used, especially in cruisers, so as to obtain large arcs of training, and also the benefit of the central-pivot mountings, but unless they are high out of the water, and project only slightly from the ship's side, sponsons have decided disadvantages, except in perfectly smooth water, as the wash of the sea against them not only causes a loss in the

" *Artillery, its Progress and Present Position.*"

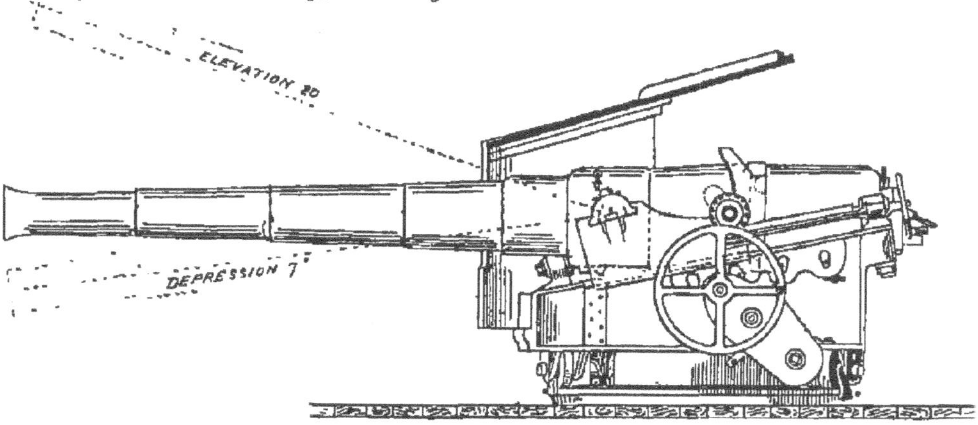

FIG. 46.—6-INCH BREECH-LOADING GUN ON VAVASSEUR C. P. MOUNTING.

speed of the ship, but often drenches the guns and everything near them, rendering it inconvenient to work them under such conditions. In the first centre-pivot mountings the arrangements of recoil-press elevating gear, etc., were identical with those for broadside mountings, except that in the 6-inch and larger mountings the length of recoil was very much reduced. Hand-worked centre-pivot mountings have been made by Elswick for guns up to 10-inch calibre on the same design as for 6-inch.

Since their first introduction, several modifications have been introduced into the Vavasseur mountings. The recoil presses have all been made "pull out," and other arrangements made for holding the gun in after recoil when necessary; the valves and pistons have also been modified, the plain orifices being now made in the steel pistons and the irregular ports in the bronze valves. In the most recent pattern of mounting, the two recoil presses are constructed quite separately, the only connection between them being a pipe, so that if one cylinder receives damage, it can thus be replaced by a spare one without disturbing the other, and the supply of spare parts is simplified.

Another improvement has been the introduction of what are known as controlling rams, which are used to regulate the speed of running out, so that it can never be violent; as with the high inclination of the slides, so as to allow of the automatic running out of the guns after recoil, even when the ship was heeling unfavourably there was always the danger, when the ship was upright, that they would run out too violently and do damage, especially if the heel of the ship favoured running out. A further modification has lately been introduced by which the action of the controlling ram can be adjusted, so that it just absorbs the running-out energy at depression, and yet does not check it too much at extreme elevation.

The system of inclining the slides, so that gravity allows the guns to run out after recoil, although possessing the advantage of simplicity is not without its drawbacks; and the most serious of these is the downward blow transmitted to the deck when the gun is fired. Rollers on a slide are a complication to be avoided if possible, but practice has shown that without them at least 10° inclination of the slides is required to insure the gun running out, even when the ship is on an even keel; but as the ship may be heeling, the slides ought to be inclined at not less than 12°. The blow transmitted to the deck when firing with a gun at high elevation, and on a slide with such great inclination, is very serious, and it therefore becomes desirable to find a system by which this blow can be avoided, and the use of powerful springs seems to overcome the difficulty, as by their use the recoil can be made to take place in the line of fire, by which actual blows on the deck can be obviated, although a heavy strain must always be communicated to it from the recoil press. Springs were first introduced into the quick-firing mountings, and the experience gained warrants their more extended employment. With a central pivot mounting, however, the strains can be distributed in a manner impossible with a broadside mounting resting on four rollers, which alone take and transmit the whole of the downward strains.

With a view of reducing the work involved in training centre pivot mountings to a minimum, a method has been found by designing the mounting so that the revolving portion rests on a ring of "live" rollers, instead of on rollers carried by brackets attached to the slides. "Live" rollers take the load on their circumference, and not on their axes; they are free to move independently of the mounting. To keep them equidistant they are attached by axial pins to a steel ring (fig. 47). The rollers so arranged run between the mounting above them and pivot plate beneath them, the actual surfaces with which they are in contact being termed respectively the "upper" and "lower" roller paths. The rollers are, moreover, provided with inner flanges, and are coned to run at the radius of the roller path without skidding. As the live roller ring is rather exposed to injury from shot, some system of protection is necessary. This may be obtained by sinking the mounting into the deck, so that the roller ring and upper roller path are below it; but a special platform has to be provided for the lower roller path or pivot plate, as the deck is not used for this purpose, the plan therefore is open to objection. The protection sought for, therefore, is usually found either by providing very stout clip plates, or by building up a glacis on the deck to surround the mounting, or by making the shield carried by the mounting low enough to protect the rollers and roller path.

The system of "live rollers" has made it practicable to

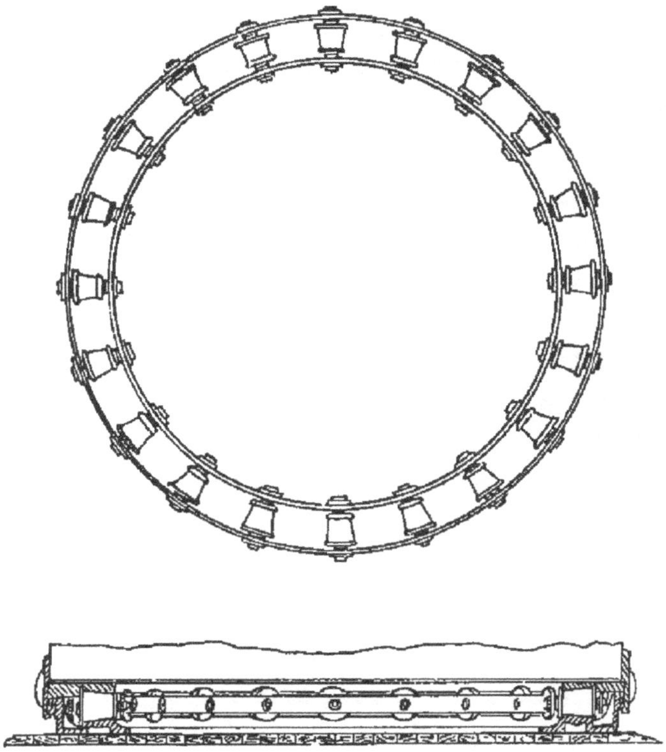

FIG. 47.—LIVE ROLLER RING.

train large mountings with heavy shields, weighing with the 9·2-inch or 10-inch guns up to 70 tons, by hand, and it is thus that it has become possible to mount such big guns in cruisers. The arrangements for absorbing recoil and elevating are practically the same both in the larger and smaller mountings. The shields for the large mountings are intended to keep out projectiles from small quick-firing guns; there is generally an inner shield or screen 3 inches thick, and an outer shield varying from $1\frac{1}{4}$ inch to 3 inches in thickness.

As showing the great strides which are being made year after year in gun-mountings as well as guns, it may be mentioned that in 1890 the largest hand-worked naval monntings were those supplied to the *Thunderer* by the Elswick firm for her 10-inch 29-ton guns, when that ship was re-armed (fig. 48); two of which were mounted in each turret, and every operation for which can be performed by hand except the training of the turrets, thus allowing the removal of the hydraulic machinery by which her old 38-ton muzzle-loading guns had been worked, by which a saving of 40 tons in weight was gained. The captain of the turret in his central sighting station, can elevate or fire either gun without assistance.

But a great advance has been made since then, and in the *Centurion*, *Barfleur*, and *Renown*, which all carry a couple of 10-inch guns in each of their two barbettes, each pair of guns can be worked by hand, although with the shields, mountings, and turn-tables, a total weight of some 226 tons has to be moved. The guns are mounted on two gravity-return slides, which are provided with hydraulic brakes for checking the recoil, and are secured to the turn-table, which is carried by a live roller ring. With these guns all the loading is done by hand. Under ordinary circumstances the training would be done by steam, the hand-gear being only brought into use in the event of a breakdown in the other machinery.

We now come to the system of hydraulic mounting of the heavier nature of guns, which is in use in all the battle-ships carrying 12-inch guns and upwards. The *Thunderer* was the first ship in which hydraulic power was first extensively used, but in her case it was experimental, being

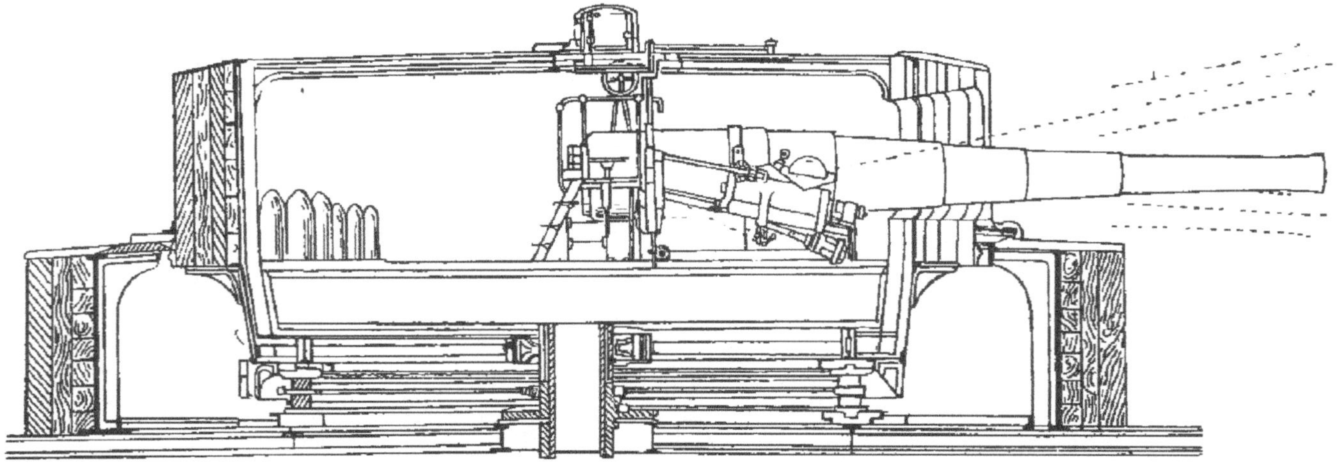

Fig. 48.—Turret of H.M.S. "Thunderer, showing Mounting of 10-inch Breech-Loading Gun.

applied, when she was first commissioned in 1878, to work the two 38-ton muzzle-loaders in her fore turret; the system employed was that which had been by this time brought to perfection by Elswick, and which proved such a decided success that every ship of a later date, where hydraulic mountings were required, have been provided with them worked on the same principle, but on a more extended scale. Having proved so successful in British ships, hydraulic mountings were soon adopted abroad, and there is now no navy where ships with large guns are not provided with hydraulic power. It would, however, appear now as if hydraulic power for the purposes of working guns has had its day, as it is being superseded by electricity; for although electric motors in the English Navy for purposes connected with gunnery are only at present being used very tentatively, yet in France in all the latest battleships, hydraulic power has been done away with entirely, and the turrets are revolved, guns elevated and depressed, and ammunition brought up from the magazines entirely by electric power; and it would seem probable that electricity, as a motive power, in our own Navy, will before long supersede hydraulic power. However, for many years past, by unanimous consent, hydraulic power has been considered the most suitable for working heavy guns at sea, and it is at present, anyway, the motive power in our ships, as there is still a certain mistrust in electricity, which is given to playing pranks, if one may use an expressive but perhaps frivolous term in speaking about such a science; and, among other disadvantages. when there is a failure in an electric installation it is not always easy to discover the cause.

The following advantages can therefore be claimed for hydraulic power for working gun-mountings, using the words of Commander Lloyd, R.N., who, as a gunnery expert holding a high position in the Elswick firm, where the system was first initiated and has since been brought to such perfection, speaks with unrivalled authority on the subject:

1. Hydraulic pipes can be "led" all over a ship without causing heat.

2. If a pipe is damaged, no explosion can take place, and it is easy to discover where the damage exists.

3. Hydraulic machines can be applied directly, as elevating presses, lifts, etc., without the interposition of gearing, and they work in perfect silence.

4. The non-elasticity of water is of particular convenience in the application of hydraulic machinery to gun-mountings.

5. The hydraulic system lends itself, in a way impossible with steam, air, or electricity, to arrangements for combining recoil presses with means for running the guns in and out.

In the hydraulic system used for working the heavy guns, and all the mechanism in connection with them, the pressure, amounting to 1,200 lb. on the square inch, is obtained by means of a force-pump, which is actuated by steam; this pump draws water from a tank built into the ship and forces it into the main-pressure pipe, from which it is led to the different parts of the mechanism. The general principle of the hydraulic system is for the pressure-water to be admitted to a cylinder by a valve worked by a hand lever, and the pressure forces out a piston, the head of which either pushes a piece of mechanism before it, or has a wire attached to it, which either raises a cage, lifts a projectile, or performs some similar work. To reverse the operation, the valve is put to exhaust and the water runs away, by an exhaust system of pipes, to the tank from which it originally came. The pressure into, and the exhaust from, the turn-tables of barbette or turret ships pass through a joint known as the centre-pivot pipe. It is a large casting divided into two compartments, one for the pressure, the other for the exhaust. These compartments have ports communicating with hollow collars which surround them, and which form the ends of the pressure and exhaust systems of the turn-tables. The collars, although arranged to be quite water-tight, are free to revolve about the central casting which, as its name implies, is placed in the exact centre of the turn-table. A tube is provided in the axis of the centre-pivot pipe for the rod controlling the training engines to pass through, in cases where these engines are not placed in the turn-tables.

Heavy guns, which have no trunnions, are secured to a cradle or carriage by the interlocking of thrust rings on the gun and on the carriage; steel bands passing over the gun and keyed to the carriage keep the guns in place. The carriage moves in and out on a slide pivoted at the front end, so that the breech can be raised or lowered by two hydraulic rams on which the slide rests. The recoil-press cylinder is placed between the slide beams and is on the push-in or compression principle, the ram or piston-rod being attached to the bottom of the carriage. But the recoil-press also provides the means of running the guns in and out.

There are eight recoil valves fitted to a mounting for a 13 5-inch 67-ton gun. Each valve has a strong spiral spring always tending to close it; it is, moreover, limited in its lift by a stop. To put the valve together, it is necessary first to thread the valve proper on the spindle until it rests on its seat. The collar, which acts as a stop, is next screwed on and pinned; then the spiral spring is placed; and, lastly, the washer and nut are put on, the latter being screwed down until the spring is compressed sufficiently to give the valve its proper load. In the case of the Elswick recoil valves, one turn of the nut corresponds to 520 lb. load per square inch on the valve. If the valve is not quite tight when closed, there will be a constant leak whenever the rear of the recoil-press is open to pressure, which means a waste of power, but if the springs are screwed down to their proper load, there is very little chance of leakage. If the springs are screwed down too much, the recoil pressure will be raised and the recoil shortened; if the springs are not screwed down sufficiently the reverse will happen. To run the gun out, pressure is admitted both to the front and to the rear of the recoil cylinder; it acts effectively on the sectional area of the ram or piston-rod. To run in, the rear is opened to exhaust; the front to pressure. Arrangements are also made for preventing the gun being run out or in violently against the front or rear buffers (fig. 49). In the accompanying plate B is the recoil cylinder; C the ram, which is connected at the front end to the horn A, projecting from the gun-carriage. The main group of recoil valves is at E, E, E, but there is a single recoil valve in the rear end of the ram C. The exhaust from this valve takes place through the orifices just in front of the piston, and keeps the space in front of the piston filled during recoil. The exhaust from the recoil valves E, E, E, passes into the main exhaust system through the pipe H, which is jointed so as to allow free movement to the rear of the slide during elevation and depression. The "disconnecting valve" is shown at G, and it is to this valve that the pressure pipe from the gun-working valve is connected. The object of the disconnecting valve, G, is to confine the pressure caused by recoil to the recoil press cylinder. F is the valve for automatically controlling the running in and out of the gun. D D is one of the cast-steel

"Artillery, its Progress and Present Position."

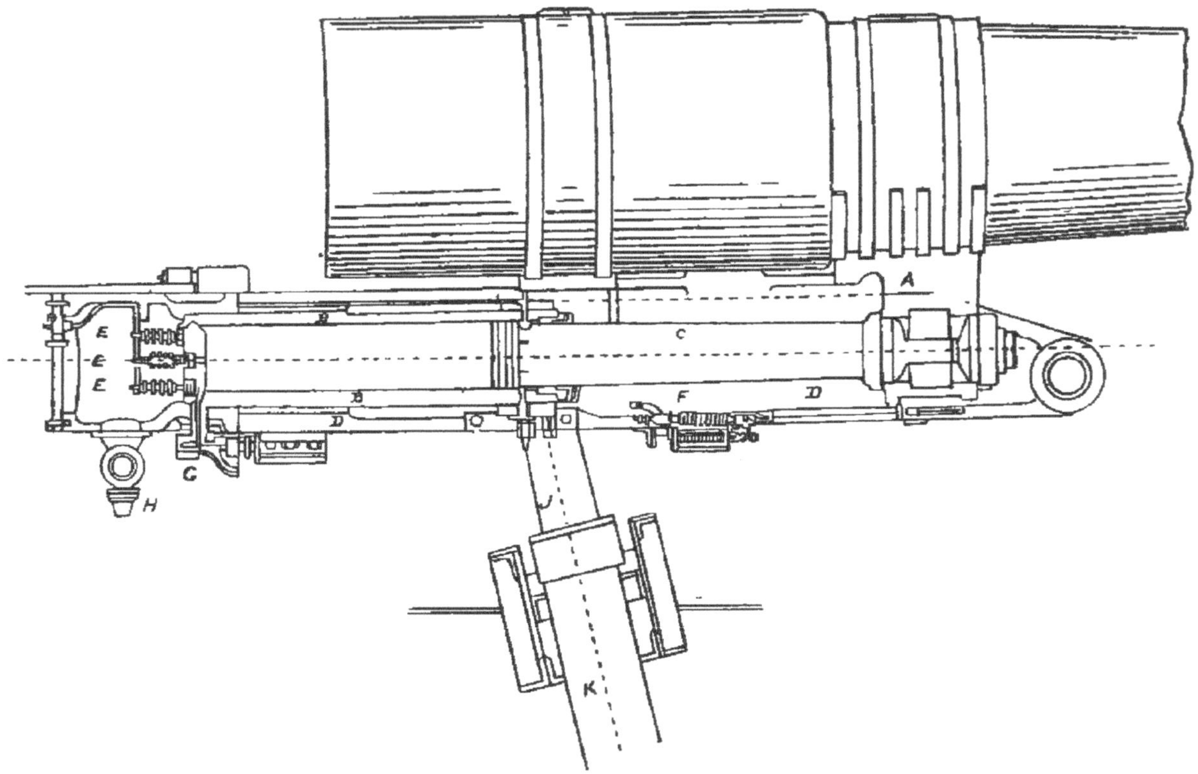

FIG. 49.—HYDRAULIC MOUNTING FOR TURRET GUN, SHOWING THE RECOIL AND ELEVATING PRESSES.

slide girders. It will be seen that underneath it is a bearing for the upper end of the connecting rod, which oscillates in the ram, J, of the elevating press, K. There is a similar elevating press under each slide girder, the two presses being supplied with pressure or opened to exhaust through one valve.

It has already been mentioned that the slides on which the carriage or cradle of the gun rests are pivoted at the extreme end, and as the centres of gravity of the guns are always in rear of this point, it is evident that the power required for raising the heavy breech end of the guns and slides, in order to give depression to the guns, must be very considerable; and in addition to the actual weight to be raised, there is the sudden shifting of the weight of the gun through a distance of about four feet during recoil. To meet these heavy strains, two elevating presses are provided, one acting directly under each slide beam, and at the most advantageous position for meeting the strains. The presses consist of cylinders of large diameter, having hollow rams working in them. A rod with hemispherical ends is placed inside the ram, the lower end resting in a socket in the bottom of the ram, and the upper end fitting into a similar socket in the under side of the slide beam. The rod is free to move and adjust its position according to the relative positions of the ram and slide, for the former moves in a definite straight line, and the latter on the circumference of a circle, the centre of which is the slide pivot. As there are no pins or bolts securing the oscillating rod either to the ram or the slide, it is a simple matter to dismount it, although there is no possibility of its becoming displaced as long as the weight of the gun and slide is upon it. Pressure is only admitted to the elevating press to give depression or to reduce the elevation of the guns. To elevate, the press is opened to exhaust, and the weight of the gun and slide is sufficient to bring them down, and to force the water out of the press. A relief valve in the supply pipe of the elevating press relieves any abnormal pressure.

Many changes have had to be made in the elevating valve-gear, in order to fully meet the difficult conditions required; since the action must be rapid, the weight on the hand levers required to work the valves must be light, and it must be possible to elevate or depress the guns from three, or sometimes four, positions, viz., from each of the two sighting stations (in some ships there are three) and from the rear, the latter position being required when laying the gun for loading.

It was very difficult at first to reconcile all the conditions, as the necessary gearing and shafting, for working the valve from so many different stations, introduced a great deal of friction, but the difficulties were met by what is known as the auxiliary elevating system. In this system large valves are only provided in the rear positions, they are furnished with long levers, and men can apply their full strength in working them. The guns are laid roughly and quickly by the rear valves; then the firer works a small "auxiliary" valve, capable only of giving a slow motion, in order to obtain the exact elevation. There are certain disadvantages, however, connected with this system; as, for instance, the man at the rear can interfere with the proper working of the auxiliary elevating valve by not closing his own large valve; then the auxiliary elevating valve necessitates a double set of valves, while it is a further disadvantage, being only able to give slow motion to the guns from the sighting stations. So, after much consideration, a new method was introduced, by which the auxiliary valve is done away with, and the man in the rear of the gun moves the same valve, by a separate lever, to bring the guns to the loading position, as the firer does to lay the guns, while the exhaust and supply valve levers are placed conveniently together, the former on the left hand of the latter. This new method apparently works in a satisfactory manner.

West & Co., Southsea, photo.

To face p. 158.

QUARTER-DECK OF H.M.S. "ROYAL SOVEREIGN" LOOKING FORWARD, SHOWING AFTER BARBETTE AND LOWER FIGHTING TOP.

CHAPTER VIII.

QUICK-FIRING GUNS.

THE French mitrailleuse, used first in the Franco-German war of 1870-71, was the weapon from which all the later machine and quick-firing guns have been developed. It consisted of a number of fixed barrels, and could fire from thirty to fifty rounds per minute. The advantages of the new weapon were soon recognized, as it was evident that the power thus given to a few men of pouring a shower of bullets on any spot might have far-reaching effects both in naval and land battles.

Shortly after the Franco-German war several inventors of machine guns appeared in the field, amongst whom the best known were Dr. Gatling and Messrs. Nordenfelt and Gardner. The Gatling gun was the first adopted in the English Navy, and it has done service in several small campaigns. It is a revolving gun; that is to say, ten barrels, fixed in frames, revolve about an axis, each barrel being fired in succession when it arrives at a certain point of the revolution. It is capable of firing some 500 rounds a minute, but, unfortunately, it is very liable to jam, especially if worked by inexperienced men; and it is somewhat difficult to keep the gun steady, owing to the handle by which the mechanism is worked being on one side. The Gatling was superseded by the Gardner gun in the Navy. In the Gardner the barrels are fixed and he side by side horizontally. Volleys or single rounds can be fired according to the rapidity with which the crank handle is revolved. These guns were made with one, two, or five barrels; they can fire about 120 rounds per barrel per minute.

The Gardner, however, in its turn has had to give way to the Nordenfelt, which took its place, not only in our own Navy, but in nearly every other service in the world, as the most reliable mitrailleuse yet constructed. The gun has five barrels of rifle calibre fixed side by side, like the Gardner, but it is actuated by a lever working backwards and forwards; also, like the Gardner, it can fire volleys or single rounds. The mechanism is simple and strong, and the action very reliable. The rate of firing is about 600 per minute. But this gun has now been superseded by the 3- and 6-pounder quick-firing guns.

The most ingenious development of machine gun is the automatic gun invented by Mr. Maxim, which, with one barrel only, can fire with most marvellous rapidity, without any exertion on the part of the firer, and without any of the difficulties in keeping the gun steady, such as are found in other machine guns, on account of the force necessary to work the mechanism being applied on one side. The gun consists of two parts, viz., the recoiling and non-recoiling. The recoiling portion is the barrel and its extension, which latter contains the lock and the gear connected therewith. The barrel is an ordinary rifle barrel with a breech sleeve, to which is secured the extension of the barrel; this extension is of steel and consists of two vertical plates. The lock is secured by a divided screw to the connecting rod of a crank, the shaft of which passes through the plates; there is also a handle to the shaft on the right of the gun outside, marked F on fig 51, and a curved arm to the crank marked B on the same figure. The lock does not differ materially from that of an ordinary pistol.

The non-recoiling portion of the gun consists of the gun-case and the water jacket, which are dovetailed together. On the right side of the gun-case is a solidly attached

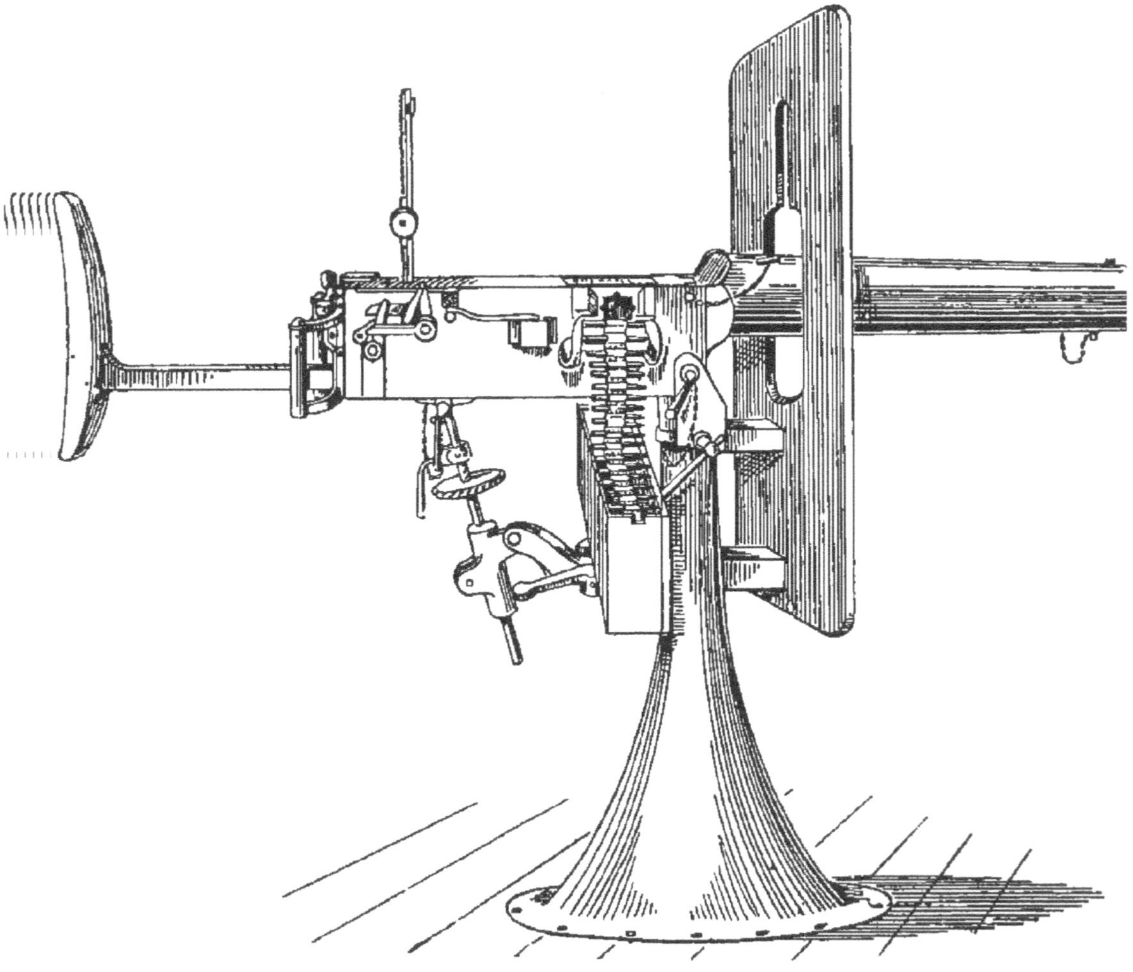

FIG. 50.—MAXIM GUN ON NAVAL MOUNTING.

resistance piece, marked C in fig 51. On the left side of the gun-case there is a strong spiral spring, the rear end of which is connected by a chain and fusee with the crank shaft, and the fore end is attached to the gun-case by means

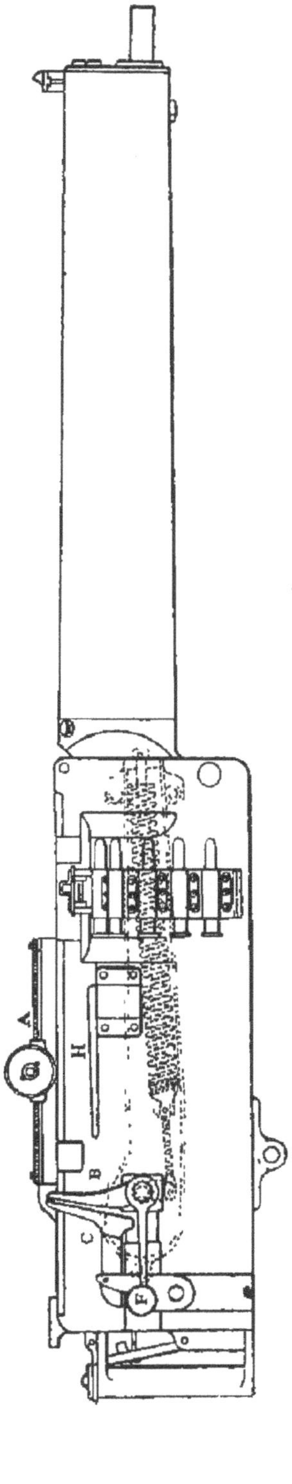

FIG. 51.—MAXIM GUN.
The Spring for bringing Barrel back, after Recoil, and Working Crank is shown, but it is really Inside, and not Visible from the Exterior

of the spring case. The use of this spring is to bring back the barrel, after the recoil, and to work the crank; the working strength of the spring can be adjusted by means of a screw on the fore end. The water jacket is of gun-metal and surrounds the barrel, it is fitted with three openings, one for receiving the water, one for drawing it off, and the third for letting off the steam, for the heat engendered in the barrel when firing rapidly speedily boils the water in the jacket. The gun is supplied with cartridges from a belt which passes through the feed-box on the top of the gun from right to left. In the feed-box are two movable pawls and two stationary ones. The movable pawls are connected by a lever to the barrel, and are adjusted in such a manner that the barrel on recoiling moves them from left to right; by means of a spring they engage behind the next cartridge in the belt, and thus the cartridges move on automatically towards the chamber. When the barrel returns after the recoil, the pawls place the cartridge, still in the belt, immediately above the chamber. The gun is fired by pressing a firing button in the rear, the recoil of the barrel setting the automatic loading arrangements in gear. If the pressure is taken off the firing button immediately after the explosion of a cartridge, the gun will be automatically loaded and will remain ready for firing, thus single shots may be discharged; but if the pressure is kept on the firing button, the gun will not only be loaded automatically but also fired, and this will continue, the rapidity of fire varying according to the energy of recoil and the length of cartridge; the average rapidity being from 600 to 650 rounds per minute. These guns just described are commonly known as machine guns, and we now come to the quick-firing gun proper.

Quick-firing guns ought rather to be called quick-loading guns, because it is in the method of loading that the great advantage of the system lies, brass cartridge-cases being used instead of silk cloth. Although heavier than cartridges in silk cloth, yet the brass cases have the advantage of being cartridges and obturators at the same time, and of allowing a gun to be reloaded with a fresh charge immediately after firing, as sponging or washing out can be omitted without danger; and even with guns up to 6-inch calibre the cases are not so heavy as to interfere with one man's power of lifting the charge; and they can, moreover, do duty as a rammer for pushing the projectile home, a duty which certainly could not be performed by means of a silk-cloth cartridge. In the 3- and 6-pounders and smaller natures of quick-firing ordnance, the powder and projectile are in one case, and the term "simultaneous" loading is used in regard to them, as against "separate" loading, used in the larger quick-firing guns, where there are two operations, the projectile being entered first, and then the powder in its case. In the matter of rapidity of fire, as the result of experiment it has been found there is but little difference between simultaneous and separate loading in quick-firing guns; on the other hand, separate loading very much lightens and accelerates the work of ammunition supply, for it renders it possible to keep a considerable quantity of shot and shell in racks round the gun, so that it is only necessary to hoist the cartridge cases up from the magazine. But the condition, which has decided the question in favour of separate loading, is the serious objection of keeping loaded and fused shell in the same magazine with large charges of explosive.

As early as 1877 Mr. Nordenfelt produced a 1-inch gun, designed for use against torpedo-boats, and worked with a breech mechanism and lever in the same manner as his rifle calibre gun already mentioned. With four barrels it throws projectiles weighing 7 drachms at a rate of about 216 a minute, and was considered powerful enough to stop any torpedo-boat at that time, and, accordingly, for many years it was supplied to all vessels as an anti-torpedo-boat gun, wherever it was possible to provide positions for it.

But while England adopted Nordenfelt guns in the first instance for repelling torpedo-boat attacks, France and other countries adopted a revolving gun, the invention of Mr. Hotchkiss. Mr. Hotchkiss constructed various revolving guns, but his 37-mm. (1·46-inch) calibre gun met with most approval; the projectile being over one pound in weight could be used as a shell, and it was no doubt considered that shell would be more generally effective against torpedo-boats than solid projectiles.

Torpedo-boats, however, increased in size, and their boilers were arranged so as to be protected by coal, so it became necessary to provide guns which would discharge heavier projectiles with a high velocity, but with the same rapidity of fire. Both Mr. Nordenfelt and Mr. Hotchkiss devoted themselves to the task, with the result that they produced the first quick-firing guns throwing 3-lb. and 6-lb. projectiles, which fulfilled all the necessary conditions; the guns of both makers being equally well adapted for the purpose in view.

From the 3- and 6-pounders from twenty-five to thirty unaimed shots can be fired by a trained man per minute, and fifteen rounds have been fired at a target in a minute, twelve out of the fifteen being hits. When it is remembered that a torpedo-boat must be repelled or destroyed before she can get within 400 yards of the ship, and that she can cover 1,000 yards or even more in a couple of minutes, it is evident that a ship cannot have too many of these small quick-firing guns for her defence, especially as she may be attacked by the boats from different quarters at the same time; these guns are therefore placed wherever there is room for them, and in such positions that several can be concentrated on any point, and they have superseded the large machine guns as having far superior range and effect, besides possessing greater accuracy.

Both the Nordenfelt and Hotchkiss 6-pounders are supplied to the Navy, as the same ammunition can be used for both descriptions of guns, but the 3-pounders are all Hotchkiss.

The Hotchkiss 3- or 6-pounder gun is made up entirely of steel, in two principal parts, the barrel and the jacket, the latter being shrunk on the former, the two being locked together by a ring or hoop screwing on to both. The breech of the gun on the exterior is enlarged from the rear of the powder-chamber, and somewhat squared, in order to give a large bearing surface to the breech-block. The jacket carries the breech mechanism, the breech-block belonging to what is known as the wedge system; it is worked by a two-handled lever on the right of the gun (fig. 52), its distinctive feature being that it works vertically, and in opening is allowed to fall freely. The mechanism of the breech-block is very simple, the various functions required of it being:—(*a*) to open and close the breech; (*b*) to fire the guns; (*c*) to extract the empty cartridge case. The block moves vertically, and falls of its own weight; therefore the mechanism for opening and closing has to lift the block straight up in closing, and to keep it either from falling open when closed or from falling out of the gun when open. It is lifted by means of a crank having a stud on the end, which works in a slot-way on the right side of the block, the block resting entirely upon it; when, therefore, the crank is turned, the stud sweeps down or up through its arc and the block is carried down or up with it. A stop-bolt screwed through the left side of the breech keeps the block from falling out of the gun. When the breech is closed the block is kept from falling down by the crank being slightly past the vertical to the front, and also by a light automatic spring catch on the handle. To load, one man opens the breech by turning the lever from right to left, and another pushes the cartridge into the chamber. The breech is closed again by the reverse motion to opening, the lock in the breech-block being cocked by the action. Loading in no way interferes with the firer, who can all the time be keeping his sights on the object to be fired at, elevating or training by means of the shoulder-piece, which is one of the distinctive features of Mr. Hotchkiss's guns (fig. 53).

The gun is fired by a trigger, there being a hammer precisely similar in shape and action to the hammer of a sporting rifle. The hammer is pivoted on an axle, called the rocking-shaft, which is hung to the lower front part of the breech-block; on the end of this shaft, projecting from the right side of the breech, is a short lever or cock-

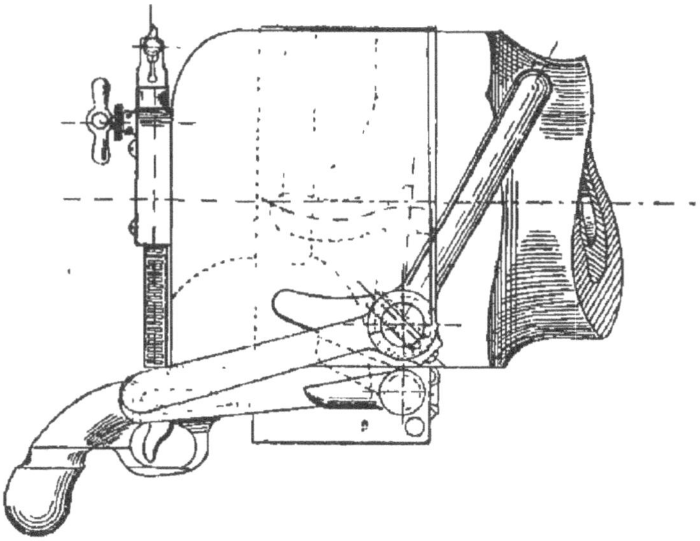

Longitudinal Elevation.

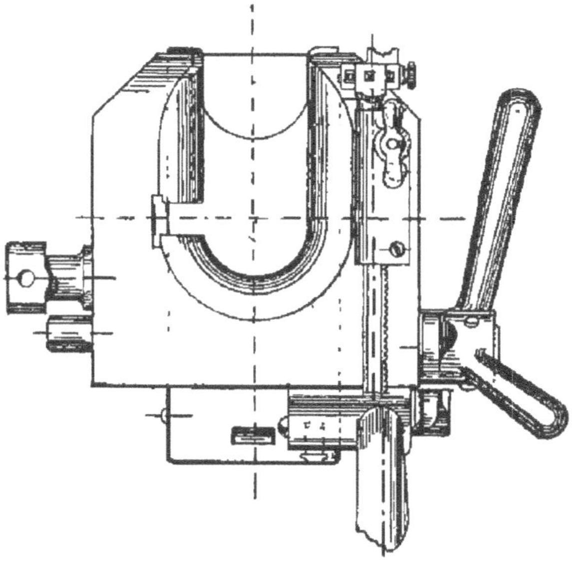

Rear Elevation.

FIG. 52.—BREECH MECHANISM OF HOTCHKISS 3-POUNDER QUICK-FIRING GUN.

ing-arm, and on the crank axle is another short lever or cocking toe. When the crank is turned, in opening the breech, the toe presses on the arm and thus revolves the rocking-shaft, throwing the hammer back.

The empty cartridge cases are extracted by a small bar or steel with a stout claw on the end, which engages in a slot-way in the breech-block, so that the down-and-up movement of the block gives an out-and-in movement to the extractor; the empty cases are thrown well to the rear of the gun. Only percussion firing is used, electric firing not having been found desirable in so small a gun.

It is to Elswick again that the Navy is indebted for the mountings of these guns, as, after a competitive trial in which several inventors submitted designs, the Admiralty selected the Armstrong recoil mounting for the 3-pounder, and finally adopted the same mounting for the 6-pounder.

As it was contemplated to use these guns for field service, trunnions were essential, with the result, however, that the mountings were not quite as compact as they might have been with trunnionless guns, but the mounting as devised has been thoroughly successful and has stood the test of many years, not one single instance of failure has been recorded. The mounting is made up of three different parts: the upper carriage, the revolving bracket, and the pivot (fig. 53). The upper carriage is one gun-metal casting, consisting of the frame and four cylinders, one cylinder to absorb the recoil and another to contain the spiral running-out spring being arranged on each side of the gun. The trunnions of the gun are fixed into "trunnion boxes," which are connected in front to the recoil pistons, and in rear to the springs. The recoil cylinders are bored out with a slight taper, so that although there is considerable clearance round the pistons when at the front end of the cylinders, the former fit the latter at the rear end. Thus, when the gun commences to recoil, the liquid can pass freely from one side of the piston to the other, but as the piston travels to the rear the clearance gradually diminishes until the energy of recoil is absorbed, a fairly constant pressure being maintained throughout recoil. A connecting channel between the two cylinders made through the casting insures the pressure being equal in both.

The upper carriage is also fitted with trunnions, and it is

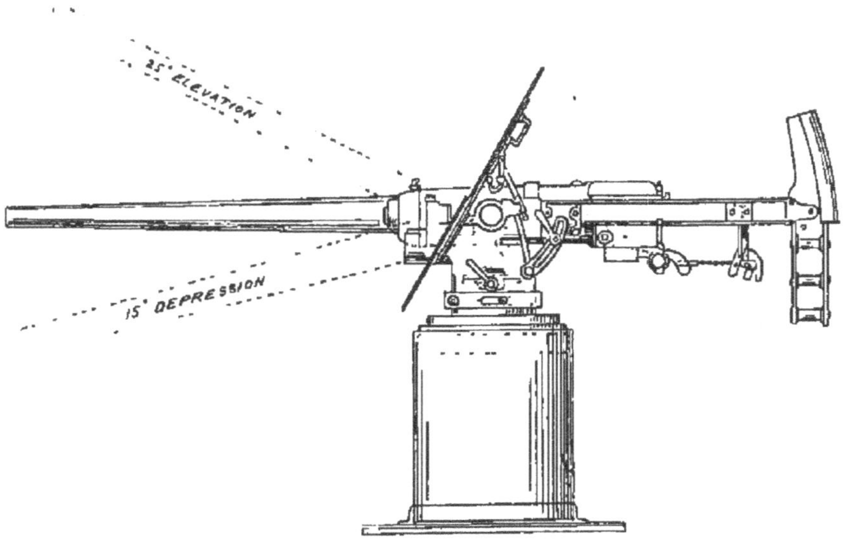

FIG. 53.—6-POUNDER HOTCHKISS QUICK-FIRING GUN ON ELSWICK AUTOMATIC RECOIL MOUNTING.

about these that the gun and carriage oscillate when elevation is given; the trunnions on the gun being only used as a connection to the carriage. The shoulder-piece is attached to the carriage (on the left side), thus none of the concussion or motion due to the recoil of the gun is felt by the firer.

The brackets supporting the carriage trunnions are in one steel casting, which fits on a pin rising from the pivot; by this means rotation for training may be obtained. A gun-metal ring, put in place in two halves, and fitting over flanges on the revolving bracket and on the pivot, forms the "clip ring."

The recoil mounting for the 6-pounder differs slightly from that for the 3-pounder, because the Admiralty desired to have a mounting suitable for the Nordenfelt guns as well as for the Hotchkiss gnus, and these have different external dimensions. The spring cylinder in the 6-pounder mounting are underneath the recoil cylinders instead of directly in rear. To save space there is also a different arrangement of gland for the piston-rods.

Most foreign countries have adopted for their navies a small single-barrelled Hotchkiss quick-firing gun, having a calibre of 37 mm., which fires a projectile varying from 1 lb. to 1·5 lb. in weight. These guns are mounted in the tops of ships and used in their boats; in the English Navy, however, the smallest gun mounted in the fighting tops is the 3-pounder.

A few years elapsed before any further advance was made in constructing quick-firing guns of larger calibre than the 6-pounders, but experiments were being carried on, and in 1886 Elswick again carried off the palm, when, after a long trial, their 40-pounder quick-firing gun, which was eventually designated and is now known as the 4·7-inch quick-firing gun, was finally adopted into the service by the Admiralty in November, 1887. This gun forms the principal armament of a large number of our second- and third-class cruisers and also the secondary battery of the first-class battleships *Nile, Trafalgar, Centurion,* and *Barfleur.*

During the Portsmouth experiments a very interesting trial was carried out in order to compare the rates of firing of the new gun and of the 5-inch breech-loading gun. Both guns were mounted in gunboats, and fired under precisely similar conditions, with the result that ten rounds were fired from the 4·7-inch gun in 47 seconds, while it took 5

West & Co., Southsea, photo.

To face p. 170.

6-POUNDER QUICK-FIRING GUN. NO. 1 LAYING THE GUN.

minutes 7 seconds, to fire the same number of rounds from the 5-inch gun. These results were very remarkable, and now that the quick-firing mechanism has been successfully applied to 6-inch and 8-inch guns, the introduction of the system has necessitated a serious modification in the designs of modern warships, to which attention will be drawn in another chapter. The 4·7-inch gun being an established success,[1] Sir W. Armstrong and Co. were soon in the field again with a quick-firing gun of 5·5 inch calibre, firing a seventy-pound projectile. This gun passed through its trial so well at Portsmouth, in 1887, that the manufacture of a 6-inch gun, throwing a projectile of 100 lb., was undertaken. The first of these guns was submitted for trial when ordinary powder only was in use, and to obtain a high velocity a charge of no less than 55 lb., contained in a cartridge case nearly 38 inches long, was necessary; such a long and heavy cartridge made loading somewhat troublesome, while the dense smoke from it made quick aiming impossible, unless there was sufficient wind to blow the smoke away.

But although the first 6-inch gun was quite successful, the Admiralty did not immediately adopt it, although the Italian and some other governments did so; the *Piemonte*, already alluded to, carried six of these guns, in addition to six 4·7-inch.

In 1890 the new smokeless explosive, "cordite," was, however, so far perfected that its use in quick-firing guns could be safely counted upon. The English service charge of this material for a 6-inch gun only weighs 15 lb., and gives a much higher velocity than the old charge of brown

[1] It may be interesting to note that the first ship to be commissioned fully equipped and armed with quick-firing guns was the *Piemonte*, a fast cruiser built for the Italian government in 1888 at Elswick.

prismatic powder; moreover, the necessary length of the cartridge case for the cordite charge is only 23 inches, so the inconvenience of the long and heavy cartridge case was overcome, as well as the difficulty in aiming the gun rapidly on account of smoke.

The Admiralty therefore decided on adopting the 6-inch quick-firing gun, and in October, 1890, some important trials were carried out at Silloth, the Elswick artillery range in Cumberland, with a gun and mounting made at Elswick to Admiralty order. The following were some of the principal results:

Five rounds were first fired for rapidity, without aiming, with a charge of 34 lb. of E. X. E. powder and a common shell weighing 100 lb. The total time was 61 seconds, the men working the gun being untrained, and having gone through the operations of loading, etc., only a very few times before the commencement of firing.

Five rounds were next fired under similar conditions, but using 15 lb. of cordite instead of powder. After the third round firing had to be stopped for a few minutes, but the first three rounds were fired in 24 seconds, and the last two in 15 seconds. Smoke of a reddish-brown colour was given off by the cordite, but it was so transparent that it would never have interfered with laying the gun, and it rapidly disappeared, although there was no wind.

Five rounds were next fired with 34 lb. of E. X. E. powder and common shell weighing 100 lb., for rapidity, and aiming at a target 900 yards off, which consisted of two 36-gallon casks, lashed together, with a red flag on a pole between them, the whole resting on the sand. The results were recorded as follows:—first round, 10 yards short of target; second round, 10 yards over target; third round, hit target; fourth round, hit target; fifth round, 10 yards over target. Time occupied, 61 seconds.

The wind had freshened, but there was still much smoke, which obscured the target. The next five rounds were fired under similar conditions, 15 lb. of cordite being substitnted for the powder charge. The following results were obtained :—first round, hit target; second round, hit target; third round, hit target; fourth round, hit target; fifth round, 5 yards short. Time occupied being only 55 seconds. The casks supporting flag were entirely destroyed and the flag perforated.

Five rounds were next fired, again with powder charge, but the fire was changed from one target to another (three targets being laid out at ranges of 900 yards, 1,400 yards, and 1,800 yards, and spread out so that the gun had to be traversed through a considerable arc). Two trial rounds were first fired at second and third targets to obtain elevation. The results of the five rounds were as follows :—first ronnd, 900 yards range, hit target; second round, the same; third round, 1,400 yards range, 50 yards over and 1 yard left; fourth round, 1,400 yards range, hit target; fifth round, 1,800 yards range, hit target, cutting flagstaff. Time for rapidity not counted, as each ronnd had to be carefully aimed and considerable delay was caused by the thick smoke, which moved slowly up the range.

Five rounds were then fired at the targets under similar conditions, but using cordite, with the following results :—first round, 900 yards, hit target; second ronnd, 900 yards, hit target, cutting flagstaff; third round, 1,400 yards, 10 yards over; fourth round, 1,400 yards, 5 yards over; fifth round, 1,400 yards (the 1,800 yards target not being visible), 20 yards over.

Time occnpied, 1 minute, 37 seconds.

The cartridge cases were extremely good, and were extracted without the slightest difficulty. Further trials were afterwards carried out, which resulted in the decision to reduce the diameter and capacity of the chamber. This alteration increased the facilities of loading, as it enabled the breech-screw to be reduced in diameter and weight, and the powder-chamber being brought to nearly the same diameter as the bore, a better "run home" is obtained for the projectile.

Before the gun, however, was finally adopted, it was sent to Portsmouth and mounted in one of the gunboats, when another series of exhaustive trials were carried out On one occasion ten rounds were fired in 1 minute and 31 seconds, with the gunboat rolling heavily; but good as this was considered at the time, it has been much exceeded at the prize-firing target practice on board some of the ships in which these guns have been mounted, where fifteen rounds with twelve hits have been fired in 1 minute.

The gun and mounting having been thoroughly tested to the satisfaction of the Admiralty, it now forms the main armament of all our first-class cruisers, and the secondary battery of the first-class battleships. It has also largely been adopted in foreign navies (fig. 54).

The dimensions of the different quick-firing guns are as follows:—The new wire 4·7-inch gun weighs 42 cwt., has a total length of 16 feet 2 inches, with a length of bore of 40 calibres, including chamber; the rifling consists of 22 grooves of polygroove E. O. C. section, ·5 inch wide and ·04 inch deep, the twist increases from 1 in 100 at breech to 1 in 34 35 at 6·65 inches from the muzzle, the remainder being an uniform twist at the latter pitch. The gun is wired for half its length; there is first the steel A-tube, then the wire, next the B-tube, extending to muzzle, and the jacket. The cordite charge is 5 lb. 7 oz., and the projectile is 45 lb. in weight; the muzzle velocity is 2,188 feet-seconds, giving a penetrating power of 7·8 inches of wrought iron at 1,000 yards, and the average number of rounds which can be fired per minute is ten. The new 6-inch wire gun weighs 7 tons 13 cwt., has a total length of 20 feet 9 inches, with a length of bore of 40 calibres, including chamber; the rifling consists of 24 grooves of the polygroove hook section ·6 inch wide and ·05 inch deep, with a twist increasing from 1 in 60 at breech to 1 in 30 at muzzle. It is constructed as follows :—The A tube, over

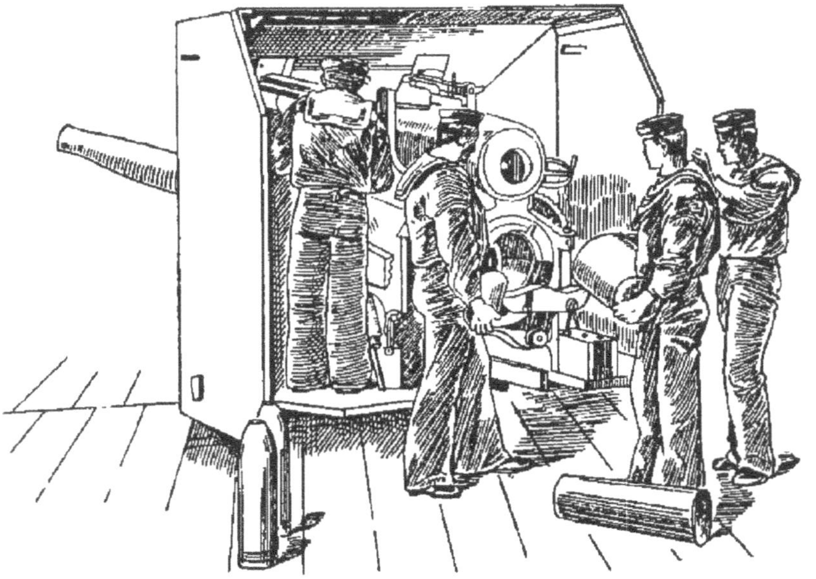

FIG. 54.—6-INCH ELSWICK QUICK-FIRING GUN, BREECH OPEN, READY FOR LOADING.

which is shrunk the 1-B tube, secured longitudinally by means of shoulders on the A tube and a steel bush at the rear, which screws into the 1-B tube, the bush being threaded internally to receive the breech-screw; round the 1-B tube is wound the wire for half the length of the gun, there being from 14 to 32 layers of wire (fig. 55) ; then comes the B hoop, which is shrunk over portions of the wire, the 1-B tube and the A tube; next the 2-B tube is shrunk round the A tube at muzzle and partially overlaps the B hoop, finally, over the exterior of the wire, the B hoop, and a portion of the 2-B tube, is shrunk the jacket, secured longitudinally by means of shoulders on the B hoop and 2-B tube, and the breech-ring which is shrunk over a portion of the steel bush, and screwed while hot to the jacket at the rear The gun fires a cordite charge of 13 lb. 4 oz., and a 100 lb. projectile, which has a muzzle velocity of 2,200 feet-seconds, capable of piercing 10·9 inches of wrought iron at 1,000 yards.

The other new wire quick-firing guns are the 4-inch, which throws a 25 lb. projectile, and with a muzzle velocity

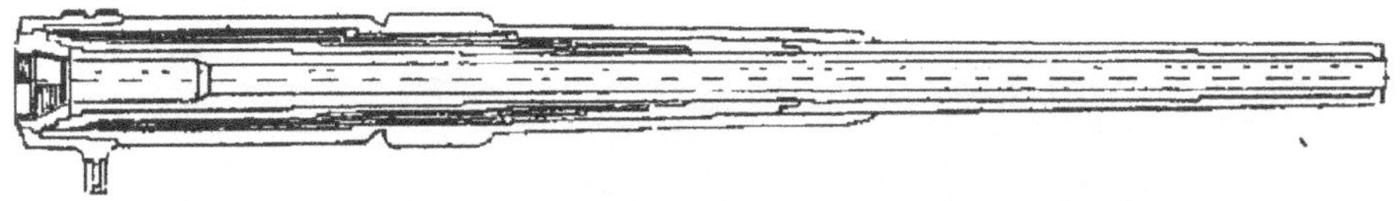

FIG. 55.—SECTION OF 6-INCH QUICK-FIRING WIRE GUNS.

of 2,456 feet will penetrate a 7 8-inch plate at 1,000 yards, and the 12-pounder gun, which in the larger cruisers is taking the place of the 6-pounder quick-firing gun, and one of which is also mounted forward in all the new torpedo-boat destroyers. The weight of the 4-inch quick-firing gun is 26 cwt., it is 13 feet 9·6 inches long, and the length of the bore is 40 calibres, including chamber. The 12-pounder weighs 12 cwt., with a total length of 10 feet 3·6 inches, and a length of bore of 40 calibres, including chamber; but there is a lighter nature of this gun which only weighs 8 cwt., is 7 feet 3 inches long, and is designed for use as a boat or field gun, but it has only a muzzle velocity of 1,600 feet, as against 2,200 of the longer type.

West & Co., Southsea, photo.

To face p. 176.

6-INCH QUICK-FIRING GUN ON UPPER DECK, SECURED INBOARD

CHAPTER IX.

QUICK-FIRING GUNS (CONTINUED).

All quick-firing guns are mounted so that they recoil in the line of fire, and to carry out this principle they are constructed without trunnions, but with longitudinal keys, or guides, on the top and bottom of the jacket, which fitting into keyways, or grooves, in the cradles, prevent the guns from revolving about their own axes. Over the rear part of the gun is shrunk a breech-ring, forged with a horn projecting downwards for attaching the gun to the oil buffer, and it also carries the hinge for the breech-screw on the right. In actual construction, except with the wire guns, quick-firing guns are built up as are other guns, the hoops being connected either by screwing or by serrations. The most important feature in connection with quick-firing guns is the Elswick coned breech-screw, which is on the principle of the interrupted screw, but is partly made of a coned instead of a cylindrical shape. The Elswick authorities claim that two advantages are secured by this arrangement; firstly, the action of opening and closing the breech is much simplified, as the withdrawal and bringing away of the breech-screw can be done in one motion instead of two; and, secondly, the coned shape enables the screw not only to take hold of the inner surface of the breech hoop or jacket, but to distribute the engagement; and therefore the strain and support, over a much larger portion of the transverse section of the gun. The thread of the parallel part has three interruptions, similar to those of the breech-screws belonging to the smaller calibres of breech-loading guns; the thread of the taper portion is also interrupted in a similar manner, but the divisions in relief are placed opposite the plain portions of the parallel part; thus the strain is distributed throughout the entire circumference of the breech-screw, instead of, as formerly, half the circumference being lost by the interrupted spaces. The breech opening of the gun being prepared in a corresponding manner admits of the screw being locked by the sixth of a turn. In order to facilitate the opening and closing of the breech, a steel tappet-ring is fitted to the outer face of the breech-screw by an end-plate, in such a manner as to admit of its being swung through part of a circle before the breech-screw is started, the momentum thus obtained materially assisting the unlocking and locking of the screw. The carrier is only an arm of steel hinged by a bolt to the right side of the breech ring; a projection formed on its inner side enters a corresponding recess in the breech-screw, and a spring clip serves to retain the screw in position when the breech is open; this clip is automatically disengaged from the screw in closing the breech.

To secure the breech-screw in its locked position, the cam portion of the lever falls into a recess in the gun when the breech is closed, and thus prevents any movement during firing. In the smaller quick-firing guns, such as the three and six pounders the extractor is made to entirely eject the old cartridge case, and no harm is done by its falling freely to the ground, but with the larger guns this would not work, as the heavy case would not only crush a man's toes if it fell on them, but it would be knocked out of shape if it struck the hard ground or the deck.

In the large quick-firing guns, therefore, there are two motions for the extraction of the cartridges. They are first started by a powerful extractor, which only has sufficient motion to insure their being free for the remainder of the extraction, the conical shape of the cartridge and chamber rendering a small motion sufficient for this purpose, and they are then completely withdrawn and laid on the ground by means of a hand extractor. The mechanical extractor is worked by the carrier in opening the breech-screw; it consists of a steel spindle passing through one side of the gun at the breech, and fitting into the groove for the rim of the cartridge case, in such a manner that when it is turned about its own axis, the fitted part acts as a lever and prises the cartridge to the rear. The extractor is brought back into its place as the breech is closed by means of a strong spiral spring outside the gun, which also serves as a buffer to prevent the breech-screw and carrier being swung too violently round The extractor is fitted either on the right-hand side or the top of the chamber, so that it is out of the way of loading, or damage from the projectile, when the latter is being entered.

In all quick-firing guns great importance attaches to the firing gear, which must be capable of being used either with electrical or percussion primers, and the change from one system to the other must not occupy any time. Electric firing has the advantage of being safer than any method requiring concussion or friction to put it into operation, as it can only be set in action in a special way. Loaded cartridges with their electric primers fitted to them, and therefore completely ready to be put into the gun, can without any objection be stored in the magazine; on the other hand, cartridge cases with percussion primers could never be stowed there, as the explosion of one charge would be so serious, that the risk, however unlikely an accidental explosion might seem, would never be run. Then, again, there is no possibility of accidentally firing a cartridge fitted with an electric primer by rough handling during loading, but an extra-sensitive percussion primer might, under such circumstances, be a source of great danger.

The firing mechanism is so arranged that the gun cannot be fired until the breech-screw is locked in the gun, and the cam lever partially lowered. A striker containing an insulated steel pin is carried in the axis of the breech-screw, and is kept pressed forward (against the primer if the gun is loaded) by means of a very strong spiral spring (fig. 56). A steel link gears with the breech-screw cam lever, and with the striker, in such a manner that if the lever is raised the striker is withdrawn, and the spring compressed. It is impossible, therefore, for the insulated pin to make contact with the primer unless the lever is "down," and as the latter can only be put down when the breech is closed and locked, the gun cannot be fired except under those conditions; thus safety against firing until the breech is thoroughly screwed up is absolutely secured. The firing battery is attached to the mounting (fig. 57), and is connected by means of a cable to the insulated contact of an electric lever fitted to a bracket, attached to the rear end of the cradle; on the end-plate of the breech-screw is another insulated contact, connected by an insulated electric wire with the outer end of the needle in the striker; when the gun is out, the contact on the lever is opposite the contact on the breech-screw, and on the firing trigger being pulled, the circuit is completed by the lever touching the contact on the breech-screw. Should the electric circuits, and there are two of them, be shot away, percussion firing can be resorted to. A percussion tube is placed into an adapter, which is screwed into the base of

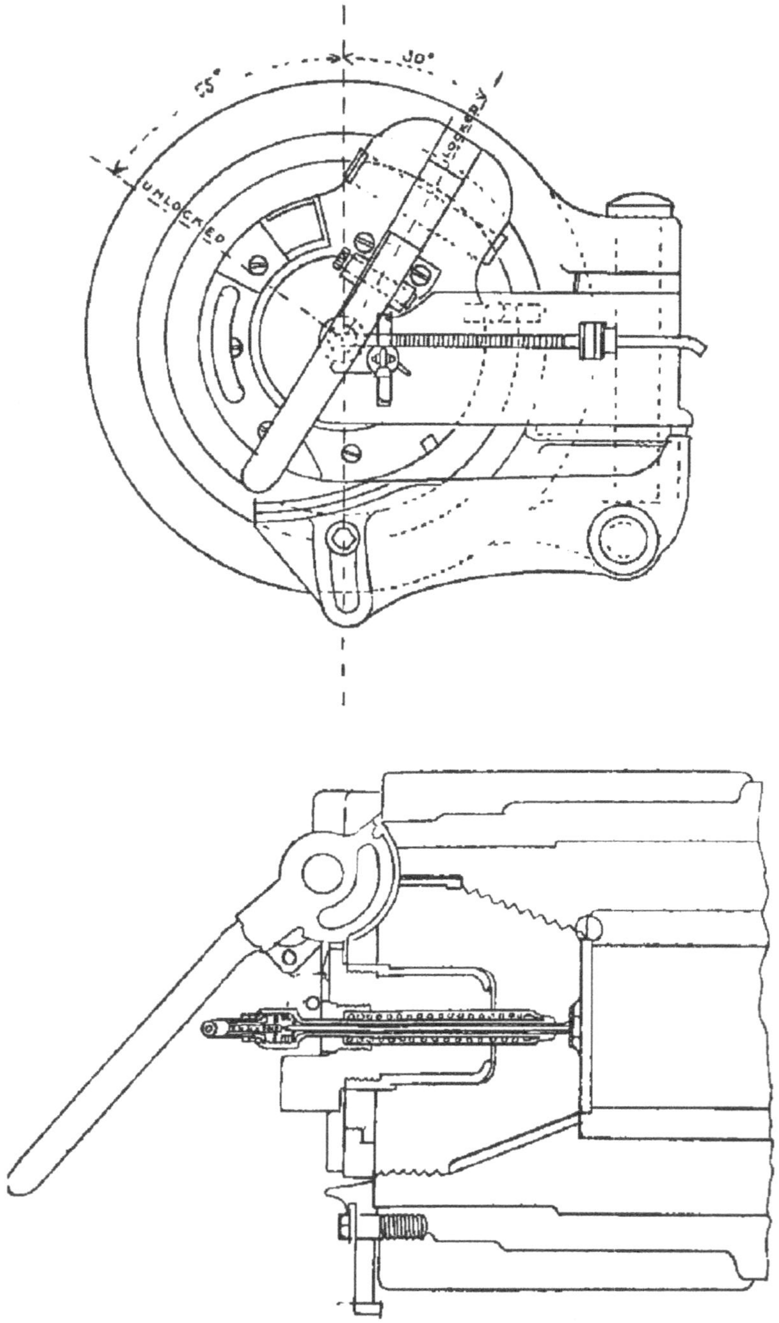

FIG. 56.—ELECTRICAL AND PERCUSSION FIRING GEAR FOR QUICK-FIRING GUNS.

(From "Artillery, its Progress and Present Position.")

the cartridge case in place of the electric primer; on one side of the striker, and fitted to the carrier, is a trigger, arranged so that it can in a moment be put in gear for percussion firing or out of gear for electric firing. If in gear, the trigger engages and the striker in the breech-block withdraws to the rear when the cam lever is raised, at the same time compressing a strong spiral spring, which, on the captain of the gun pulling on the tube lanyard, when the lever is down again, is released, causing the striker to fly

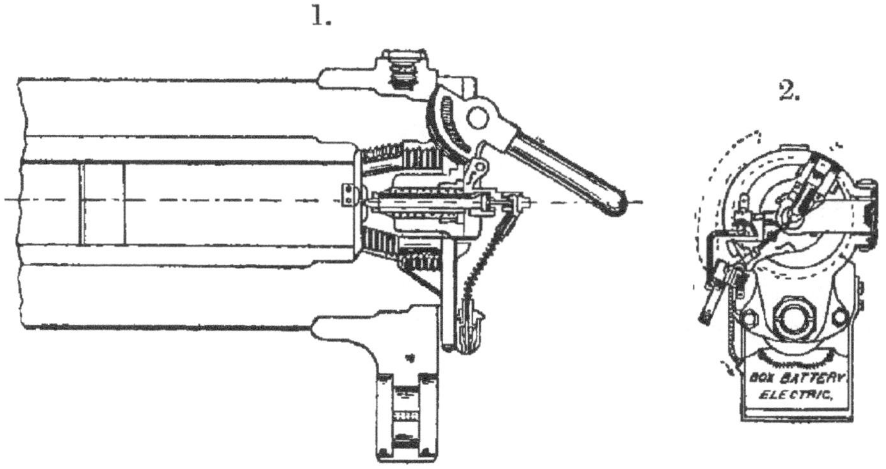

FIG. 57.—4·7-INCH QUICK-FIRING GUN.
Showing (1) Vertical Section of Breech, and (2) Rear View.

forward, against the cap in the centre of the tube and so firing the charge.

It has already been mentioned that quick-firing guns recoil in the line of fire, and to obtain this recoil in the line of fire the gun is slipped into a cradle that acts as a guide for it, a horn connecting it (as has been already mentioned) with the pistons, which work in the recoil-press cylinders under the cradle. The recoil energy having been absorbed, the gun is forced out into the firing position by the running-out springs, violence being checked by the controlling ram. The spiral steel springs used for running out are of very special manufacture, and are so excellent that there has never been an instance of their breaking on service, although the number of rounds fired from quick-firing guns using such springs must be numbered by hundreds of thousands.

It has been found difficult to find room under the cradles to give sufficient length for the springs, and they have therefore been placed on top, where, although they are more exposed, they are more accessible. In the 6-inch guns, in order to save space and yet to get as much effect as possible out of the springs, they are arranged in two sets, one set fitting inside the other; and both inner and outer springs are in two lengths, so that, should a spring fail, only a small portion would require replacing. In the 6-inch and pedestal mountings, the springs are in a separate containing case, the cover of which is attached to the top of the breech-ring, and this case moves with the gun during recoil. The springs take their bearing in front against the end of the case, and in rear against a steel plate or cap; a rod, secured to a bracket on the fore end of the cradle, passes through the springs, so that, on firing, the recoil of the gun draws the plate into the spring case like a piston, and compresses the springs, which, as soon as the recoil ceases, force the gun out again. To give an idea of the work done by springs, it may be mentioned that those for the 6-inch quick-firing guns give a resistance of 7,400 lb. at the commencement of recoil, and 13,800 lb. at the end. The springs for 4·7-inch guns give an initial resistance of 3,000 lb., and of 5,800 lb. at the end.[1]

In the ordinary 4·7-inch quick-firing mountings the trunnions of the cradle rest on cast-steel brackets, which stand upon, and are riveted to, the upper roller path, which is shaped by hydraulic pressure out of a solid plate of forged

[1] Lloyd and Hadcock, "Artillery, its Progress and Present Position."

steel. A 3-inch forged steel shield is also bolted on to the roller path and to the brackets. By this means the carriage is supported and strengthened, besides being protected. The upper roller path runs on twenty-four steel rollers, which are made with interior flanges, and are kept in place by the roller ring to which their axes are attached. The pivot plate or lower roller path is of cast steel, the path being accurately machined. With live rollers, as has been already stated, friction is reduced to an absolute minimum ; so the guns can always be trained without gearing, which is, however, provided for use if required. The 6-inch quick-firing mountings are somewhat differently constructed, although they are on the same principle The side brackets are cast in one piece with the upper roller path. To give solidarity, as well as to afford protection to the mounting and gun's crew, a 3-inch steel shield, forming a nearly complete cylinder, is bolted on to and covers the cast-steel carriage.

Quick-firing mountings below the 6-inch are designed to train by a shoulder-piece if desired, but the 6-inch mounting is considered rather too heavy for this, but with the worm gearing disconnected, one or two men can push the mounting round, or train it by tackle. In all but the pedestal mountings clip plates are fitted to take the lifting strain on the front when the gun is fired, and on the rear in case the gun runs out too violently.

The gun is elevated by means of a hand-wheel which gears with shafting into the elevating arc on the cradle. This hand-wheel is conveniently placed for use with the left hand of the man who is laying the gun, who can, at the same time, keep his right eye on the sights, have the firing pistol in his right hand, so that with his left shoulder pressed against the shoulder-piece, he can train, elevate and fire the gun without moving his eye from the sights. With the 6-inch gun, which cannot be trained with a shoulder-piece, it is necessary to train with the right hand, and then move it to the firing pistol to fire.

On account of their length, it has been found necessary to arrange for the running in and complete housing of the new guns, which are mounted between decks; if this were

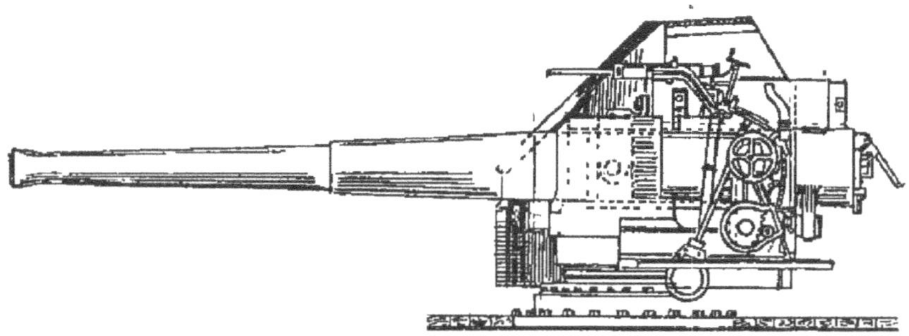

Upper-deck Mounting.

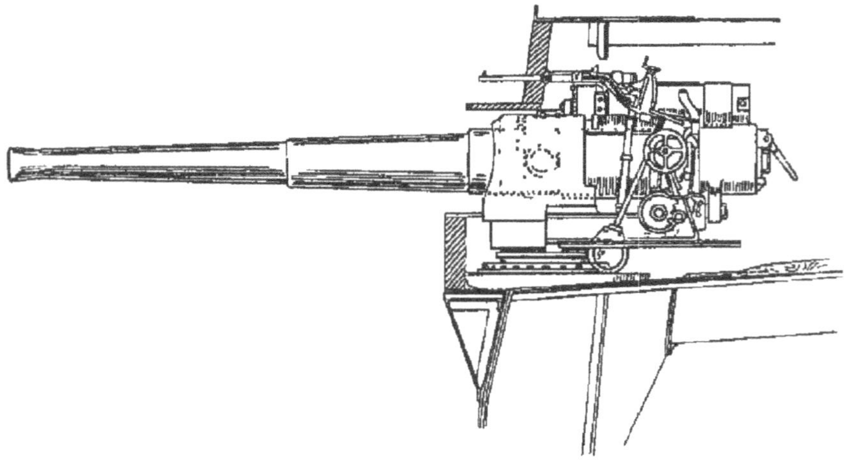

Between-deck Mounting.

FIG. 58.—6-INCH QUICK-FIRING GUN ON CENTRE PIVOT PEDESTAL.

(*From "Artillery, its Progress and Present Position."*)

not done it would be impossible for ships to go alongside each other or jetties; and, moreover, if the guns were kept out at sea, their muzzles would suffer very much from rust, and it is moreover impossible to make ports close tightly over a gun and be really water-tight. In English ships, therefore, all quick-firing guns mounted between decks are fitted with arrangements, so that they can be run in quickly

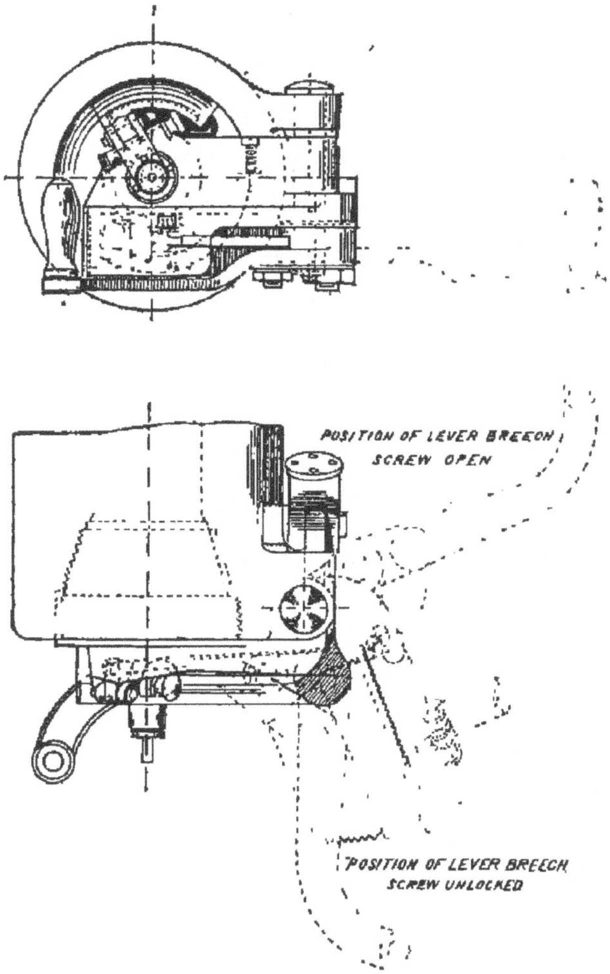

FIG. 59.—ELSWICK NEW MECHANISM FOR BREECH-LOADING ORDNANCE.

and easily; and under ordinary conditions at sea they are stowed entirely inboard, but they can be remounted ready for action in a very few minutes. The principle adopted is to hang the guns in their cradles to an overhead tramway, and then run them back. This is a very easy matter with the 4·7-inch gun, as the gun and cradle weigh under three tons; but the 6-inch gun with its cradle weighs over eight tons, and special precautions have to be adopted in order to move so heavy a weight suspended from the deck with safety, even when the ship has motion. The trunnions of the cradle fit into blocks, which slide on to and lock into guides on the carriage brackets. It is only necessary to take the weight of the gun and cradle, and not actually to lift them, when it is required to disengage them from the carriage. Six-inch guns can be housed in four minutes, starting from the firing position on the beam and can be replaced in three and a half minutes.

Elswick has lately perfected a system of opening the breech by one motion only. The hand-lever, instead of being on the top and working in an approximately vertical position, is on the lower side of the breech-screw, and works always in a horizontal plane; it pivots on the carrier, and is attached to a sliding block by means of a connecting rod. A pin in the breech-screw works in a vertical slot in the sliding block, so that a horizontal motion of the latter causes the breech-screw to turn. The centres, about which the gear works, are on their dead points when the screw is closed, and it is therefore perfectly locked. The lever, if swung round, first unscrews and then brings away the breech-screw. The coned screw is particularly well adapted to this system, for there are only two motions for it to pass through, and these are readily combined so as to give the operator but one. With the parallel screw a more complicated lever must be used, as there are three motions to be combined in one. This system has now been adopted for all the new wire quick-firing guns up to the 6-inch inclusive.

CHAPTER X.

MAGAZINES, SHELL-ROOMS, AMMUNITION SUPPLY TO GUNS, AND LOADING ARRANGEMENTS.

THERE is no point of more importance in the internal arrangements of a ship than that for the stowage of the ammunition, or in other words than the magazines. However carefully powder is manufactured, it is certain to become irregular in action if it is exposed to damp or heat; and yet it is almost impossible to fulfil the other necessary conditions of keeping the magazines on board ship closed, and of having them as far as possible out of danger, by constructing them below the water-line, without the heat in them becoming very great, and the atmosphere more or less moist. In many of our ships of war some of the magazines are built on the middle line, between the boilers, and nothing can make such positions really suitable from the gunnery point of view.

Ships' magazines are constructed as tanks of steel within the vessel, and are fitted with water-tight doors and special flooding arrangements, so that they can be filled with water in case of there being any danger from fire. They are lined with teak, but sufficient space is left between the steel plates and the teak to allow a certain amount of ventilation, and also, in case of fire, to guard against the danger of the steel plates, if they become hot, setting fire to the teak. No iron is allowed inside a magazine, all the metal fittings and tools being of copper, some alloy of that metal, or of zinc. A ventilator, in or near the floor of the magazine, with an uptake in the crown, is always fitted, in addition to the flooding and draining arrangements already mentioned. The magazines are lit by means of strong glass bull's-eyes, which are illuminated by electricity in light-boxes, only accessible from outside the magazine; candles being provided for the light-boxes as an alternative in case the dynamo breaks down. Great care and attention is required to prevent the temperature of the magazines becoming excessive, and thus causing the deterioration of the powder or cordite stowed therein, which, if then used, would cause excessive pressures to be set up in the guns and would be liable to burst them. To guard against this, ventilating arrangements are fitted between decks, consisting, in ships of the *Royal Sovereign* class, for instance, of ten fans worked by steam, and fed by cowls leading down from the upper deck; they are arranged along the ship in such a manner that the three foremost and after ones are connected together, and the four remaining work in two pairs amidships; each set can be worked in conjunction or singly if there should be any break-down. The air is forced along shafts which are tapped to admit it to the magazine. A regular weekly record is kept of the temperatures of all magazines, and those in which cordite is stowed are never allowed to be over 100° Fahrenheit.

All magazines open out of the magazine chambers into handing-rooms, separated by a wooden door from the magazines, and this door, the only entrance, should be closed when the magazines are being worked. Holes in the door, fitted with fearnought screens, enable the full and empty cartridge cases to be passed out and into the magazines without the danger of sparks or fire finding their way through. The handing-rooms are generally in communication with the deck above by means of scuttles. It is not always possible to keep the door between the handing-room and the magazine closed where heavy portable magazine cases have to be passed out and in, as the passing-scuttles are not made sufficiently large to allow cylindrical cases to pass through them.

The powder charges are kept in the magazine in hermetically sealed cases, which are of various patterns and sizes, and they are now almost invariably made of corrugated brass, which combines strength and lightness, although there are some still made of wood. It is necessary that they should be capable of being easily opened or closed, should be perfectly air-tight, and water-tight under a pressure of from 10 to 20 lb. per square inch; but they must also be as light as possible, and their form convenient for stowage. Magazine cases may be divided into two classes—portable and semi-portable. In the first class are the cases for charges of heavy guns, which are taken from the magazines, and raised to the gun with the powder still in them, and also the cartridge cases employed with quick-firing guns. Semi-portable cases are only removed from the magazine when the powder is being shipped or disembarked. The cartridges are taken out of these cases, which generally hold several cartridges, and are passed up to the gun through the handing-room scuttles in cartridge cases. The portable cases never hold more than one charge, and in the case of heavy charges, only half a full charge; for large guns these cases are cylindrical and are fitted with removable ends, so that a rammer may enter at one end and force the cartridge out at the other. The method of securing the covers to insure air-tightness, as well as to give facility for removing them, has only been arrived at after a great deal of experience and trouble. Specially prepared india-rubber packing, called "Dermatine," has given the best results, and with it there ought to be no sticking of the cover to the case, as often happened with plain india-rubber packing. Each end of the cylindrical case is provided with four lugs, the covers having four handles to correspond with them, which, when screwed down, draw the cover into place and at the same time compress the packing. The cylindrical magazine cases (fig. 60) are lifted by an eye placed in such a position that they will hang horizontally from it. For heavy guns of 12-inch calibre and otherwise, each case generally con-

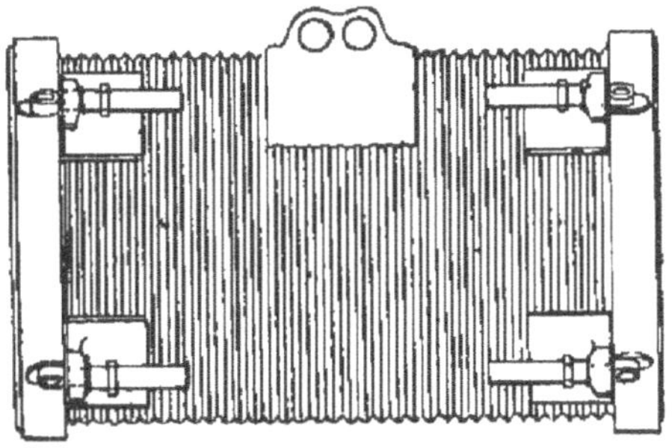

FIG. 60.—CYLINDRICAL CARTRIDGE CASE.

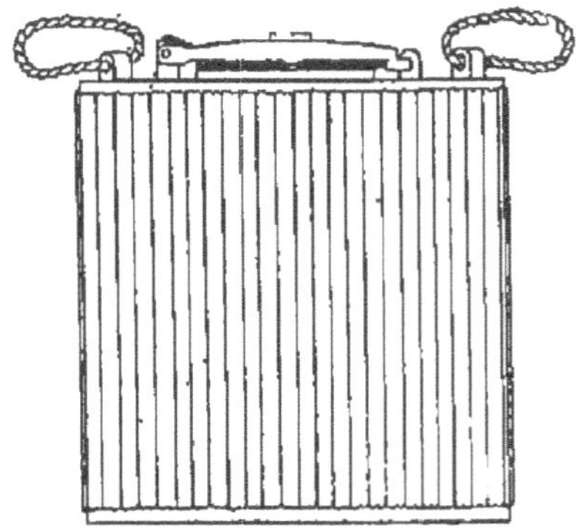

FIG. 61.—"A" CARTRIDGE CASE.

tains two quarter charges. Semi-portable magazine cases are generally rectangular (fig. 61) and are also corrugated. They open at one end, having a cover which is made water-tight with what is termed "luting," a mixture of beeswax and tallow, and are stowed in the magazine with the opening of the case outward to the passage. The utmost care should always be taken to re-seal any magazine case that has been opened but not emptied, nor ought more cases to be opened than is absolutely necessary. The magazine cases are stowed in racks, technically termed "bays," with passages running between them for convenience in working the magazine, and where portable cases are used, overhead tramways are fitted, having "travellers," with purchases running in them, for lifting the cases, the falls being taken to hydraulic bollards; as soon as the cases are lifted out of the rack, they are traversed outside of the magazine chamber, into the handing-room, the covers removed, and the cases placed in the *loading tray*. A hand rammer is used to force the powder into the lift, which, when loaded, is raised to the barbette.

In large battleships, like the *Royal Sovereign* and her sisters, for instance, there are two main magazines for supplying the 13·5-inch breech-loading barbette guns; two more for supplying the 6-inch quick-firing guns on upper and main decks; two 6 and 3-pounder quick-firing for the smaller quick-firing guns, and one machine gun magazine for the small-arm and pistol ammunition. The main magazines are situated below the barbettes; the 6-inch, 6 and 3-pounder magazines and shell-rooms all open into an ammunition passage, which runs fore and aft, amidships, below the protective deck. The charges and projectiles are whipped up into this passage by endless whips, and then carried to the different supply trunks leading to the main and upper deck, up to which they are hoisted by whip and winches to the guns. The advantages for separate loading in quick-firing guns can best be appreciated when the question of stowing and working in magazines is considered. The short cartridge cases of English guns occupy but little room, and are light and convenient to handle. If the shot were added, those for the 6-inch guns would be too heavy and bulky for hand working.

The main shell-rooms open on to the opposite end of the lift, to which the powder is supplied, and the operation of placing a projectile and charge in the lift is carried out simultaneously. The shells are stowed in long "bays" or racks, divided into compartments, which are severally devoted to the different classes of projectiles, and are lifted from their racks by means of purchases from overhead travellers, and then generally lowered on to little carriages (called bogies), which travel on rails and convey them to the position for charging the cage in which they are conveyed up the ammunition shaft to the turret or redoubt; these shafts, as has been already mentioned, are, in modern battleships, generally vertical, and lead direct from the magazine and shell-room to the gun. The cage is in three compartments; the upper carries the projectile, and the centre and lower the two half charges, and when the cages are up, one of these compartments is in line with the loading-tray and gun, if the turret or barbette is in the loading position. The cage is raised by means of an hydraulic purchase, automatic arrangements being fitted, which insure the cage being brought gently to rest in the several positions it has to occupy. In loading, the rammer passes through the cage when driving the projectile and powder into the gun; special care, therefore, has to be taken that the cage must on no account be moved while the rammer is being used. The rammer, of which, as yet, no mention has been made, is, for 12-inch guns and upwards, hydraulic, and, although protected by the armour-plating, is, except in the act of loading, outside the turret or barbette, and is fixed, which entails, except in the latest battleships, the guns being always brought back after firing to one loading position. In order to economize space, the rammer is almost invariably made with tubes, one within the other, usually with only two such tubes or rods, but sometimes with three or even four. All the moving rods are made from drawn manganese bronze pipes. They must be no larger in diameter than is necessary to give sufficient rigidity and strength to support the weight of the head when the rammer is extended (fig. 62).

On the end of each rod is formed a piston, and on the other end a gland. To extend the rammer, pressure is admitted to the rear of the fixed portion; it acts on the whole area of the piston formed on the end of the first rod, and passing through a small non-return valve, also acts on the second rod. Both rods move out together until the first reaches the limit of its stroke, and then the second rod leaves the first, and continues to extend until its piston comes in contact with the gland at the outer end of the first rod, or until the projectile or powder charge is

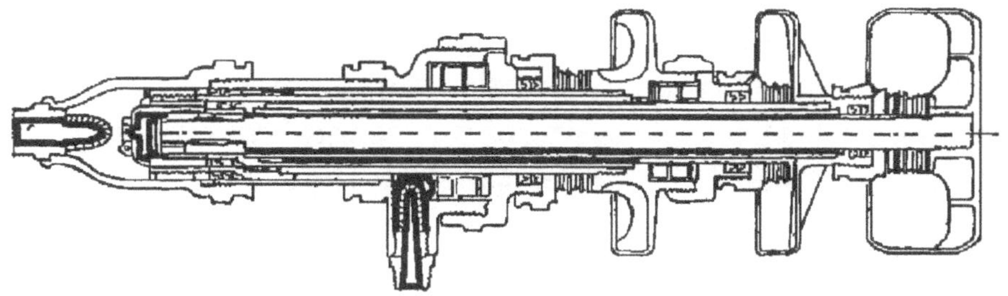

FIG. 62.—HYDRAULIC RAMMER (THREE-THROW).

home. To withdraw the rammer, the rear end is opened to exhaust; and pressure, passing through a port in the front end of the fixed portion of the rammer, acts on the front side of the piston of the first rod; it also passes up a number of small channels, formed in the thickness of the metal of the first rod, through ports into the front end of the first rod, and acts on the front side of the second or smaller rods. The rods are thus withdrawn into each other, and the rammer closes up, as it extended, like a telescope. As the rammer is made entirely of bronze or gun-metal, great care has to be taken to prevent dirt or grit capable of scoring the soft metal getting into it.

An indicator is fitted to show the position of the rammer-head, for the man attending the valve cannot see how far his rammer is extended. In ramming home it is best to let the projectile go full speed for the last few feet of travel, so as to insure its being well home, for the only thing which prevents its sliding back down the inclination of thirteen degrees, is the fact of its copper driving-band being driven into the rifling of the gun.

In a previous chapter it has been explained how the turret or barbette has to be trained to the loading position after firing, and there locked, and how, during training, the guns are elevated, breech-screws unscrewed, withdrawn, and traversed to one side, and as soon as the guns are washed out, and the loading tray placed, the signal is made to the men outside to load. While the turret or barbette is being trained, the ammunition cage, previously charged, is raised to the first position for "shot," and directly the signal to "load" is made, the following operations take place: the rammer comes forward and rams the projectile into the gun; it is then withdrawn, and the cage raised to second position, when the first half charge is rammed home, and the rammer again withdrawn; then the cage is raised to third position, the second half charge rammed home, and the rammer finally withdrawn. The signal is then made, from outside to inside, that the gun is loaded. The locking bolts are then lifted; at the same time the loading tray is withdrawn, a tube and primer placed in the vent of the breech-block and the breech-block traversed, inserted, and screwed up, after which the gun is run out and laid approximately horizontal, while the turret is being trained for the object, after which the guns are laid accurately ready for again firing. It might well be supposed from the above description that the operations of training a barbette or turret after firing to the loading position, the loading itself, and the subsequent operations before the gun would be ready for firing again, are both complicated and slow, but as a matter of fact they are neither. In a ship of the *Royal Sovereign* class, after some months in commission, two rounds have been fired at a target in one minute and forty seconds, while two rounds from the 111-ton guns of the *Benbow* or *Sans Pareil* have been fired in two minutes and thirty seconds.

In the new battleships of the *Majestic* class, which carry the 46-ton wire gun as their main armament, many improvements in the supply of ammunition and the loading arrangements have been effected. Hitherto guns in the Navy worked by hydraulic power have been loaded in the "run in" position, but in the *Majestic* and her sisters the guns are loaded when run out. Much more room is thus obtained in rear of the guns for loading operations. In the *Royal Sovereign* class, and in the earlier ships, there is, as has already been stated, but the one fixed position for loading, so that if one gun only has been fired, the other has to be practically put out of action for about two minutes until the first has been re-loaded. In the *Majestic* and *Magnificent* however, in addition to the fixed ammunition hoist, which has been retained with the idea that to load both guns simultaneously it is the quickest system, there is also a central or all-round loading hoist for the supply of the powder, while a considerable stock of projectiles stowed in the gun-house provides the other essential. Thus either gun, or both guns, can be loaded in any position, and the loading of one gun does not in any way interfere with the working or firing of the other. The guns can be loaded simultaneously at a fixed position, or separately at an all-round position, and the time necessary for working the guns has been much reduced. Although the loading at the fixed position is by hydraulic power, at the all-round position the loading has to be carried out by hand in the two ships just mentioned.

During the gun trials of the *Majestic*, the two guns in the after barbette were trained over on the port beam, and two rounds were fired against time, the fixed loading position being used for one gun, and the all-round loading position for the other. The time was taken from the firing of the first round, the assumption being that there would be ample time to load for the first round before the ship entered an action, and the most satisfactory results were obtained. In the case of loading from the fixed position, the interval between the two rounds was only one minute nineteen seconds, and in the case of loading from the all-round position it was one minute twenty-one seconds. In the case of the guns of the ships of the *Royal Sovereign* class, as we have already mentioned, the time taken to fire two rounds is a little under two minutes, the fact of their being heavier making no difference, as the hydraulic machinery is but little affected by heavy weights.

In the later ships of the *Majestic* class now being completed, an improvement has been effected by fixing an hydraulic rammer for the "all-round loading" to a floor forming part of and rotating with the turn-table. The central hoist, revolving with the turn-table, admits of the powder charges being brought up to the gun in any position of the latter, a store of projectiles being kept in the gun-house. An improvement has, moreover, been made in the method of raising the powder charges in these newer ships. A high-speed hydraulic motor (running at about 450 revolutions a minute) is fitted in the central trunk, and raises the case containing the charges from the magazine to the gun in about fifteen seconds; two cases are provided, and so arranged that one travels up while the other descends, so that in the space of about half a minute charges for both guns can be raised from the magazine.

In the 9·2-inch guns, which are mounted as the bow and

stern chasers in the first-class cruisers, the powder hoist, similarly, comes up direct from the magazines through the turn-table under the centre of the mounting. A cage is raised or lowered in the tube by means of a system of pulleys and a wire rope, which is led to a hand or steam winch at the base of the tube. Connected to the upper part of the cage, by a ball-and-socket joint, is a cylinder for the powder charge. If a door at the bottom of the tube is opened, the cylinder, no longer receiving support from the side of the tube (which alone maintains it in a vertical position) falls out to a convenient angle for charging. After the charge has been placed in the cylinder, it is lifted into a vertical position and the door closed. By means of the winch it is then raised in the tube, and when at the top its upper end protrudes into a sort of cowl or bent tube, which is attached to and rotates with the mounting. The cowl guides the powder cylinder during the remainder of its upward motion, and the ball-and-socket joint, attaching the cylinder to the cage, allows the former to follow the bend of the cowl, whatever the position of the mounting. Thus, when the cage is at its highest position, the cylinder is inclined at a convenient angle, with its mouth or upper end close to the rear of the gun, and the powder charge it contains within easy reach of the man whose duty it is to load the gun. The projectiles are not lifted up the central hoist, but they are stored in considerable numbers on deck, close to the mounting, and a derrick with a small winch is supplied for working them In the new mountings—just designed for the *Powerful*, *Terrible*, and the new first-class cruisers—arrangements are made for a store of projectiles being kept close to the gun on the mounting itself, while the motive power for raising the ammunition is electricity, a small horse-power motor being employed to do the work. At the gun trials of the

Terrible the electric hoist was found to work extremely well, as it only takes 10 seconds to bring the ammunition up from a depth of 60 feet, and 30 seconds for opening the breech and loading; whereas it takes two men, by manual labour, 32 seconds to smartly hoist up the ammunition.

In the *Centurion*, *Barfleur*, and *Renown*, three new medium-sized battleships, which mount 10-inch guns in barbettes in pairs, the ammunition supply arrangements are much the same as those already described. There is no fixed loading station, but an all-round one. In the centre of the redoubts in which the guns are mounted is a tube extending from the magazine deck to almost the under-side of the turn-table. It is divided vertically into two compartments, each of which is fitted with guides and cages. The larger compartment contains a cage capable of bringing up ammunition for the guns, and the smaller one a cage which can carry a couple of projectiles or a single charge of powder. The former is worked by a steam hoist erected on the machinery deck, and controlled from the top of the central tube, while the latter is worked entirely by hand, and is intended as an auxiliary service in case of accident to the steam gear. The ammunition, when conveyed to the top of the tube, is landed in a chamber secured below the turn-table, which revolves with it round the tube. This chamber is fitted with two hoists, one behind each gun, each capable of carrying a projectile and a couple of half charges, and delivering in rear of the breech in readiness for loading. The hoists in the chamber were originally worked by windlasses by men inside the chamber, but electro-motors have since been fitted to do the work, much quickening the transport of the charges to the gun. The ready supply of projectiles is stowed round the sides of the chamber, about sixty rounds being stowed there, which would be replenished, as necessary, from the shell-rooms, through the central tubes, either by the hand or steam cage.

It may be of interest to state that the total quantity of gun ammunition carried in the shell-rooms and magazines of the battleships of the *Majestic* class is 14,120 rounds. This is for the 12-inch, 6-inch, 12-pounder and 3-pounder guns, and is exclusive of the Maxim and rifle ammunition. Out of this amount there are between 300 and 400 rounds for each of the four 12-inch guns. The weight of projectiles amounts to, roughly speaking, 270 tons, and the quantity of cordite contained in the magazines is close upon 43 tons.

Before concluding this chapter, it may be mentioned that central-loading mountings were designed at Elswick as far back as 1881 for the 9·2-inch guns of the armoured cruiser *Impérieuse.*

CHAPTER XI.

SIGHTS, THE DIRECTOR, FIRING ARRANGEMENTS, AND CONNING TOWER.

In Chapter I. we mentioned that it was not until the end of the Great War in the beginning of the present century that proper sights for guns came generally into use, and in Chapter II. we described the improved barrel-sight introduced by the present Lord Armstrong, when his system of breech-loading guns was adopted. Since that date considerable alterations and improvements in the sights have been made.

It must be remembered that, as has been already mentioned, rifling always causes the projectile to take a path inclined slightly in the horizontal plane; a correction must, therefore, be made by the sights. The error, or "drift," is found by practice, a number of shots being fired, and the average horizontal error being taken for each nature of gun. The rear sights are inclined at an angle so as approximately to correct this error. With right-hand rifling the inclination is always to the left, the deviation of the projectile being to the right.

In the Navy what is known as the H-speed sight is the only one now used; the fore sight is a small globe, or spherical bead, on an upright thin piece of steel (fig. 64), and the rear sight consists of a fine aluminium wire stretched horizontally between two uprights (fig. 63) on a sliding nut, thus resembling the letter H, fitted to the cross-head; below the wire is a semi-bead to give the direction of the line of sight with the blade of the fore sight. Attached to the rear face of the cross-head by four small fixing screws is a removable aluminium scale-plate, graduated on its upper edge with a speed scale in knots an hour, and on its lower edge with a deflection scale giving 2° right and left. The nut is traversed by means of a screw, worked at

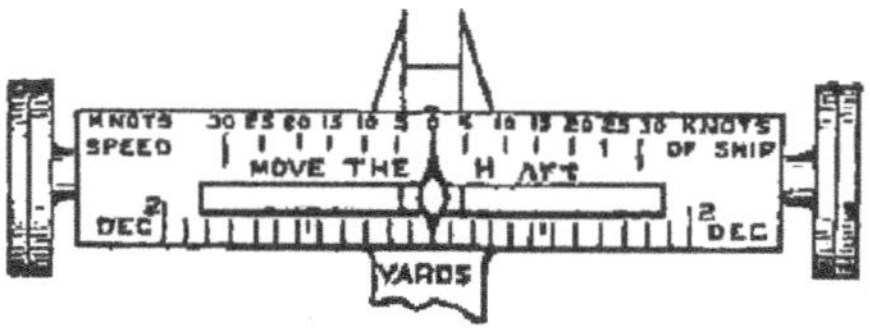

FIG. 63.—H-SPEED SIGHT.

FIG. 64.—FORE SIGHT.

FIG. 65.—SIDE VIEW.

either end by a milled head; the words "move the H aft" are engraved in red on the scale-plate, to prevent the possibility of deflection being accidentally given in the wrong direction when it is desired to allow for the speed of the ship. "For wind" the sight should be moved to windward; while "for speed of enemy" the sight should be moved in the direction of the enemy's course.

As changes in the muzzle velocity are always to be anticipated, due either to alterations in the charge or the weight of the projectile, and as such changes entail modifications to the graduation for range on the sights, the graduations are cut on strips of aluminium, which are fitted to slide into grooves cut in the sight-bars. So if it is necessary to re-graduate the sights, it can be done at the expense of the aluminium strips only; and as aluminium has a bright silvery surface, it answers the purpose admirably, and gives the figures and graduations a prominent appearance.

The method of using the sights is to make the line passing through the bead of the fore sight and the object, also pass through the middle of the wire; when the guns are laid horizontal, and the rear sight is close down, the cross-bar on the H would bisect the bead on the fore sight if you looked along them at a distant object, and the line of sight would be parallel to the axis of the gun. A ship is generally a moving platform, and the moment for firing has to be chosen when the motion brings the sight on the object. The clamp for adjusting the rear sight is what is known as Stuart's clamp, and is the invention of a Mr. Stuart, who has for many years past been in the employ of the Elswick firm. It consists of a socket containing a small hand-wheel, which carries a shaft, and can be turned in either direction, or be withdrawn rearwards in opposition to a spring. The turning movements raise or lower the sight slowly but accurately. The withdrawing movement liberates the sight-bar, so that it can be raised or lowered quickly by hand to its approximate position. The introduction of the H sights is due to Lieutenants (now Captains) Arbuthnot and Foote, who were both, as lieutenants, gunnery officers.

For turrets or barbettes mounting a pair of guns, it is necessary two sets of sights should be provided in each sighting station, and the sighting arrangements here are automatic. As the sights have to work simultaneously in the left, front, and right sighting stations, and extreme accuracy is required of them, it is not an easy matter to arrange, and gearing is unavoidable.

Attached to the lower part of the slide beams is a curved rack, which gears into a pinion carried by a shaft, which traverses the barbette underneath the gun-mountings. This shaft is also geared by pinions, one at each end, gearing in their turn into straight racks attached to standards carrying the actual sights. The sight-bars themselves slip into sockets on the standards, and are set therein to the angle of inclination required to correct drift; the sight-bars are raised and lowered in the standards by means of Stuart clamps, which have been already described. The two sets of sights, one worked by the "right," and the other by the "left," gun, are placed side by side with indices attached to them, to show when they are in exact correspondence. Should the firer wish to fire both guns simultaneously, he can thus readily observe whether they have identical elevation.

In a barbette or turret the front sight is secured to the outer part, in a line with the gun, from the sighting position. The work of training, sighting, and firing the guns (except when they are fired from the conning tower) is performed by the captain of the barbette or turret (who is a petty officer) from his sighting station, where he is provided with all the necessary appliances for directing his guns, adjusting his sights, and firing. After the guns are loaded, the bolts are raised which secure the turn-table in the barbette or turret in the loading position, and he proceeds to train towards the object by turning a wheel that is close to the sighting position in which he stands, and which sets the hydraulic turning machinery in motion, while the turret (say) turns, he roughly lays the gun for elevation, having previously set his sights for range and the speed deflection, stopping the turret, if necessary, when the sights are drawing near the object, so as accurately to adjust the elevation, the slide and consequently the breech of the gun, being raised or lowered until the holding bracket of the sight brings the bar of the H sight, the bead on the fore sight and the object all in a line with the eye of the captain of the turret, who fires as the sights pass the object. The actual firing of the charge is accomplished by means of a firing key placed close to the sighting position. This key is fitted with two terminals, one of which is connected with an electric battery, which is led to earth by a bolt in the turret, the other terminal of the key being connected by a wire to a lock, which slips into the breech-block of the gun and covers the firing tube which has been placed in the vent; when the lock is in place an insulated needle-point from it engages into some insulated white metal in the base of the tube; as soon as the captain of the turret presses down the firing key and so connects the two terminals in it, the electric current passes from the battery, through the key to the lock, along the needle-point into the insulated white metal, and along a fine piece of platinum wire, which is surrounded by some fine powder, to the body of the tube, to the gun, and so back to the other pole of the battery; as the platinum wire in the tube offers great resistance to the passage of the current, it becomes red hot and ignites the fine powder around it, which in its turn ignites a primer of cordite placed in the vent, and thus explodes the charge in the gun.

The sights are marked up to 11,500 yards, but for long ranges the H sight is removed, and a telescope fitted with cross wires is secured to a bar hinged at the fore end, at the position of the fore sight, and the rear end resting where the cross-head was; the gun is then fired when the cross wires intersect the object.

In the quick-firing guns a bead and H form of sight are attached to a bar, connected with the cradle, and the H sight is geared up to a drum on which the distance up to 10,000 yards is marked; this drum is revolved until the distance of the object comes in line with the pointer, and in so doing the H sight is raised, and when the sight is adjusted, the bead, H sight, and object are all brought into a line with the eye of the captain of the gun, who raises or lowers the breech as necessary to do so. When the gun is laid for the object, it is fired in a somewhat similar way to the heavy guns. The captain of the gun has hold of a pistol grip, and as he pulls the trigger it allows the current from an electric battery on the mounting to pass through the platinum wire bridge in the primer and so fire it, the igniter and the charge. The sights are so fitted that sighting can be done through separate ports in either the shield or ship's side, instead of through an enlarged gun port. This system allows the sights to be brought further forward, and the firer can consequently obtain a larger field of vision through the sight ports. It has been already mentioned that percussion firing can be resorted to if the electric circuits should be shot away.

In guns other than the quick-firing or barbette or turret guns the sights are fitted into sockets in the guns themselves the fore sights being placed near the trunnions, and the rear sight never being less than 36 inches from it.

Specially trained and qualified men are selected to control the working of the guns. Thus the barbettes or turrets are worked from the inside by the captain of the barbette or turret, and a second captain who superintends outside, with two captains of turret or barbette guns, one at each gun; each 4-inch quick-firing gun, and upwards, has a captain of the gun in charge, and each of the smaller guns a highly trained seaman-gunner.

An important point to be considered is the method of using guns in the dark, and for this purpose night-sights have been introduced: two patterns are in use in the service, the "Grenfell" for heavy, and the "Elswick" for the light guns.

The "Grenfell" night-sight is the invention of Captain Grenfell, R.N., and consists of two fittings, one of which can be secured to the front and the other to the rear sight. Each of these fittings is arranged to receive a small incandescent lamp, *a*, fig. 66, the light from which is reflected through a window at *b*, which is of red glass in the case of the rear sight, and of white glass in that of the front sight. This light striking on a curved surface at *c*, which is polished and of a non-corrodible metal, is seen by No. 1 of the gun as a bright horizontal line. The curved polished surface, in the case of the front sight, is confined to the central part of the stirrup as shown on the right at *d*, whilst in the case of the back sight a notch is cut in the stirrup through which the front sight can be seen, whilst on each side the light of the lamp is reflected as a red horizontal line from the two polished surfaces, *e*, *e*. The appearance in aiming is, therefore, a white horizontal line between two red ones. When the three are in one straight line and at the same time the middle line is bisected by the mark, the gun is truly aligned, both for direction and elevation. The complete apparatus is shown in fig. 67, the back being on the left and the fore sight to the right. Between them is a switchboard connecting them to the

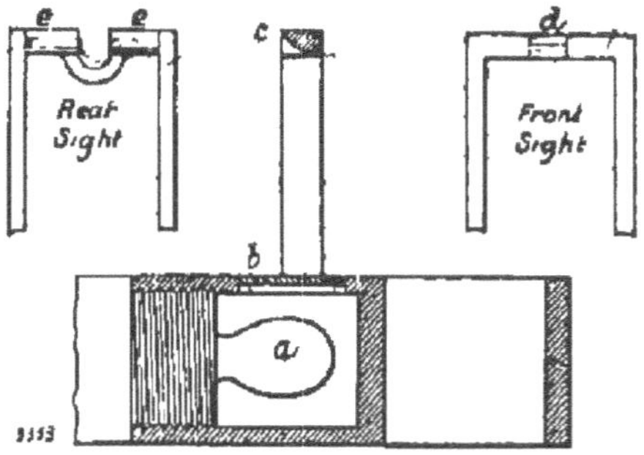

FIG. 66.—"GRENFELL" NIGHT-SIGHT.

source of current. Mounted on this switchboard are two rheostats, by means of which the intensity of the light is controlled. By operating one of these, the relative brightness of the two sights can be adjusted to each other. This done, the second rheostat is used to vary the brightness of the two lamps uniformly, until both the sights and the object aimed at can be clearly seen at the same time.

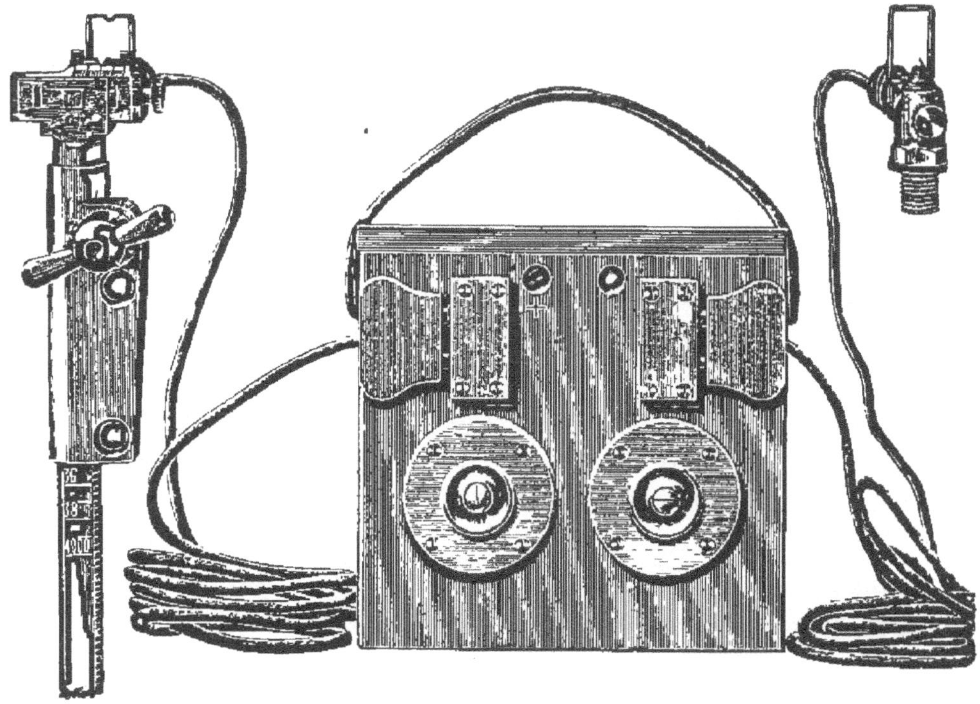

FIG. 67.—GRENFELL'S NIGHT-SIGHTS.

In a still more recent form of this sight Captain Grenfell has added a third switch, by which a small hand lamp can be coupled up with the switch. This lamp is used for reading the range scale.

In the "Elswick" pattern used for the smaller guns, the fore sight is formed by the tip of an electric light that is shipped on to the fore day-sight, and an electric lamp on the rear sight illuminates a cross wire. The captains of guns are taught to estimate the distance of objects both by day and night, and with careful training they become very proficient.

Mention has been made once or twice of an instrument known as the "Gun Director," and although in consequence of the increased rapidity of fire from turrets, barbettes, and quick-firing guns, its value has much diminished, and its use been greatly curtailed, yet a work of this kind would not be complete without some description of it.

The use of the gun director is to bring the fire from a turret, barbette, battery, or any individual gun, under the immediate control of the captain of a ship during action; and if he should so desire it, a concentrated fire from all or any of these guns could be delivered by its use at any required distance. Thus the director for the time being becomes the telescopic sight for all the guns that are fired by the firing key attached to it.

It consists of a telescope fixed by trunnions to a vertical frame, in which it is free to move up and down, and be adjusted for the necessary elevation required.

The vertical frame in which this telescope is fixed is pivoted on a graduated arc, and can be moved over it in a corresponding manner to a gun being trained over its racer. The telescope is fitted with cross wires for obtaining a fine sight.

Certain corrections are applied to the guns and director to compensate for the difference in position of the two. The reader would therefore have an accurate idea of the working of the director if he considers it as a sighting gun, which it is possible to lay for an object by looking along the axis, and giving it the required elevation and training in the usual way.

It is placed in the conning, or some specially armoured tower, from which a good all-round view can be obtained;

close to it is a firing key, fitted into a return-wire firing circuit, from which branches are led to every large gun in the ship that could deliver its fire on the broadside on which the director is fixed; these branch wires are fitted with points to which the tube that fires the gun is attached, and also a slot and bolt for connecting up when the gun is loaded, laid, and ready.

But it is necessary to more fully describe the various details and corrections that have to be applied when using the director. It is placed in the ship so that when everything is at zero the axis of the telescope is at right angles to the fore and aft line of the ship, and horizontal in all directions when the ship is on an even keel.

Converging stops are fitted for the racers of guns on the bow, beam, and quarter, to enable them to be rapidly trained into position, so as to deliver a broadside with great shattering effect on exactly the same point at 800 yards distance. The necessary elevation for the broadside is passed to it from the director (or directing gun if that system of firing is used).

Tables showing the amount to be allowed for each correction are secured to the conning tower, and in sight from the director.

These corrections consist of:

1. That due to speed. It is allowed for by means of a vernier fitted to the part of the vertical frame that travels over the arc or racer of the instrument.

2. The dip correction, which is an allowance that has to be made because the axis of the telescope is in most cases above the axis of the guns used with it; it is combined with the necessary elevation that has to be given to the gun to enable it to send the projectile the required distance, and it is adjusted by a vertical scale and vernier to the left of the telescope.

3. The elevation necessary to enable the axis of the telescope to be brought on to the object, after the "dip correction" has been applied; it is adjusted on an arc to the right of the telescope.

All these verniers, etc., are fitted with clamping screws. If the director is to be used for firing, the order is passed to all the guns in the director circuit, "Converged firing by director," followed by the bearing (bow, beam, or quarter). The speed correction is made, and the telescope clamped on the bearing given; the "dip and elevation" correction is placed on, the arc clamped, and the telescope laid, so that the cross wires intercept the spot where it is supposed the object to be fired at will pass; the elevation shown is read off and given to the guns, who lay accordingly, make "ready," and report; unless any allowance is to be made for the speed of the object aimed at, the broadside is fired by pressing down the firing key in the conning tower, immediately the cross wires intersect the object at the required spot.

Should the object not pass at the expected distance, the "dip correction" for the true distance can easily and quickly be put on; and if the ship has any motion, the cross wires will almost always cut the object aimed at at some portion of the roll; but should they not do so, the telescope and guns must be relaid for a new elevation.

There are generally two director positions, one for each side of a ship, and the guns in a barbette or turret can be fired from either; this may sometimes lead to serious accidents from the following cause. It is not unusual for the firing key to be wedged down, when testing the firing circuits, so that a tube can be fired by joining up the slot and bolt on the branch, and if, by an oversight, this wedge is not removed, a person entering the opposite conning tower, and seeing everything there all right, would proceed to lay a broadside, and immediately the turret guns are connected up, they would unexpectedly be fired by the wedged-down key in the *opposite* conning tower. It is therefore absolutely necessary to keep these instruments, and, if possible, the towers in which they are placed, under strict lock and key, and supervision of a responsible officer. In the same way an ignorant person might fire the turret guns at an unexpected moment by means of the opposite firing key, if the disengaged tower was not carefully kept locked.

The first improvement on the instrument just described is known as the "Collingwood" pattern. The main improvements consist in the trunnions of the telescope being fitted at the outer end, so as to diminish the size of the sighting port required; and the horizontal movement is controlled by a pivot under the inner end of the telescope; the arcs are done away with, and graduated movable collars round the telescope are substituted, and the speed is shown in knots; thus two sets of reference tables are done away with; but this method of applying the different corrections necessitates a director being specially graduated for its position in the ship, relative to the guns it is used with; and should it be moved to any other position, special tables would have to be prepared for use with it.

A table giving corrections for quarter and bow training is required.

The next improvement is known as the "double dial" gun director. This also has to be specially fitted to each ship. A graduated dial with pointer is placed on each side of the telescope, within easy view of the person laying the broadside; the one on the right is used for the "dip" correction and required elevation; that on the left for the training and speed correction for bow, beam, or quarter. The graduations are also carried out to a greater degree of accuracy than in the previous pattern, and the telescope is improved to give greater magnifying and focussing power.

There are two conning towers in modern battleships, of which the foremost is the principal one; in vessels of the *Royal Sovereign* class the armour of the foremost one is 16 inches thick, and of the after one 3 inches. In action the captain fights his ship from the foremost one, where he has everything connected with her under his supreme control; the navigating, gunnery, and torpedo officers are with him, a chief quartermaster at the steam wheel and compass, which are within the tower, and an orderly to attend at the voice pipes. A similar arrangement with the commander, or second in command, exists in the after conning tower. From the conning towers voice pipes lead down to the barbettes or turrets, all gun and torpedo quarters, ammunition passages, double bottom and water-tight door stations, fighting wheels between decks, signal stations and engine rooms, to which mechanical telegraphs also communicate, so that the captain can in a few seconds communicate with any part of the ship. When direct communication is interrupted, exchanges on the deck below are formed, and voice pipes lead from there to casemates, magazines, shell-rooms, etc. Electric bells and indicators are fitted to draw attention to a message being sent or received. In addition to the gun directors, the directors for the torpedo tubes are also in the conning towers, into which the firing circuits for the tubes are also led. The torpedo directors are practically the sights of the torpedo tubes, the deflection given to the torpedo on discharge by the speed of the ship, allowance for the speed of the object aimed at and tide, are made at these directors; as soon as the object aimed at is in a line with the sights of the directors, the firing key attached to it is pressed, and causes an electric current to form an electric magnet at the torpedo tube, which attracts a steel plate and frees a heavy ball; this ball in falling forces down a stop-valve, and allows the compressed air in a reservoir below the tube to escape into the rear part of the tube and propel the torpedo out.

Iron shutters are fitted between the hood of the conning tower and the armour of it, so as to give a clear all-round view when lifted.

The captain therefore sends down from the conning tower all orders to the barbettes, or turrets, and guns, as to the object to be attacked, nature of projectile to be used, distance for which fuses are to be fitted when fire is to be delivered, regulates the speed and helm, discharges the torpedoes, and receives signals from the flagships; a multiplicity of responsibility that requires minute and accurate knowledge, great nerve, and many years of experience, to bear with credit to the country and the Navy.

West & Co., Southsea, photo.

To face p. 214.

CHAPTER XII.

POWDER AND CORDITE.

In spite of the old adage, "Keep your powder dry," most people will find in these days that the subject of powder is not necessarily a dry one.

To accurately gauge the importance of the vast strides that have been made in its production, it will be interesting to note the methods of its first manufacture, and to gradually work up to the results of latter-day experiences.

It is impossible to fix the actual period at which gunpowder was discovered, but it is known that the Chinese had knowledge of its proportions and properties from very early days. It was used for defensive purposes in the seventh century, and it became known to the English, through Roger Bacon, early in the thirteenth century, but it did not become of any practical value until a German monk, named Schwartz, in 1320 introduced a systematic method of thoroughly mixing the ingredients. It was first used in England for warlike purposes in 1327, and it came into universal use by European nations a few years after this.

There is no record of any powder-mills being established in England before 1590; in 1787 the powder-mills at Waltham Abbey were purchased by the government, and in 1815 they sold the Faversham mills, where, up to that time, most of the powder had been manufactured, to a private firm. The manufacture of powder continued on its original lines, and without much improvement, until about thirty-five years ago.

English black powder is composed as follows :

Saltpetre	75 parts.
Charcoal	15 ,,
Sulphur	10 ,,

And it must be clearly understood that when these are mixed together, that they form a mechanical mixture and not a chemical compound.

The result to be aimed at is to produce a large volume of gas from the chemical combination of the constituents of the powder; the more rapidly the gas is produced, the more shattering is the effect caused by its formation. No one can state what will be the accurate chemical result of the explosion of a charge of gunpowder, but the following is a fair general idea of what happens on a charge being ignited.

As soon as the temperature of the igniting agent reaches 482° Fahr., the sulphur, which has been introduced to lower the point of ignition, bursts into flame, and the saltpetre being acted on by the heat, readily frees the oxygen with which it is highly charged, and allows it to combine with the charcoal and form carbonic acid and oxide, whilst the nitrogen is liberated.

The products of combustion are by weight about 57 per cent. of solid matter, and 43 per cent. of permanent gases.

The temperature of explosion is about 4,000° Fahr. There is about four and a half times as much carbonic acid and two and a half as much nitrogen as carbonic oxide produced by the explosion of a charge, and one gramme of powder yields about 260 c. c. of gas.

Saltpetre is found in a natural state in warm climates,

particularly in India and China, but only Indian saltpetre is used in the government factories. It is sent to England in a state known as "grough" saltpetre; and it has to be purified before it can be used for the manufacture of gunpowder. This is done by mixing the "grough" saltpetre with boiling water, and the mixture being allowed to cool, the pure saltpetre crystallizes on the surface, while the impurities either remain in solution or drop to the bottom of the liquid as sediment. It is then a simple matter to skim the saltpetre crystals off the surface of the liquid and dry them ready for use for the manufacture of powder.

Great care has to be exercised in the production of charcoal, from which the carbon is obtained, the materials used for burning are dogwood, alder, willow, and straw, according to the nature of powder required; and the temperature at which the carbon is formed has a great effect on the gases given out by the charcoal.

Sulphur is found abundantly present in various parts of the world, particularly in volcanic districts. Its presence as an ingredient in powder is not absolutely necessary, but it is used, as it ignites at a low temperature, and so facilitates the ignition of the gunpowder and helps to maintain the temperature of the resultant gases, thus adding to their volume. Like saltpetre, sulphur contains impurities, which have to be eliminated by heat before it can be used for powder, and its behaviour, when subjected to heat, is very peculiar. It melts at a temperature of 239° Fahr., but when the temperature has risen to 350° Fahr., it becomes thick again, the temperature still rising it again melts, and lastly, at about 836° Fahr., it distils as a heavy brownish-red vapour; this vapour is condensed, is run in a liquid state into a vessel, and is afterwards broken and ground up into a powder for use.

To obtain thoroughly reliable powders, all these ingredients must be of the purest possible nature. The conversion of them into gunpowder is as follows.

The sulphur and charcoal are ground into a fine powder, and the ingredients are then carefully weighed out and mixed, in charges of about 60 lb., in a revolving drum; from this the "green charge," as it is called, is taken to the incorporating mill, and in this state it is in a highly dangerous condition; the "green charge" is damped in the mill, and is crushed into "mill cake" by revolving stones. It is generally during this operation that most explosions in powder-mills occur, so special precautions are consequently taken, and, for safety, the mills are divided off into sections of six; the buildings containing the mills being constructed with substantial walls, but with very light roofs, so that an explosion would find vent by blowing the roof off; while above each mill, which is also separated from its neighbour, is a tank that would automatically drown the charge in its mill, if an explosion occurred in the building. A breaking-down machine now converts the mill cake into "meal," after which it is "pressed," then granulated to the required size and dusted. It is then glazed by placing it in a revolving drum, and the excessive moisture is extracted by placing it in a drying-room; a final glazing and blending complete the operation.

The absence of the proper amount of moisture in a powder causes very high and rapidly formed pressures to be developed on its explosion; as this would probably result in the bursting of a gun in which such a charge was used, great care has to be exercised in the stoving or drying of the powder.

In the earlier days of ordnance only small-grain powder was used, but as the guns increased in size and length, so the size of the grain also increased, until it became so large, that it was known as pebble powder, and this powder was made by the "press cake" being cut into cubes of the required size, which were completed by glazing, stoving, and being blended; as the guns continued to increase in size, this pebble powder in its turn gave way to prismatic powder, which was formed by compressing the granulated press cake, by hydraulic machinery, into six-sided prisms.

The turning-point in the history of the manufacture of gunpowder as a service store was the introduction of rifled ordnance. The main point to be borne in mind in reference to a charge is that it should produce an even, low, continuous pressure on gun and projectile, giving the latter a high velocity as it leaves the muzzle of the gun. This velocity is measured by feet per second and is known as the muzzle velocity.

Small-grain powder gave a large igniting surface, and was consumed before the projectile had almost left its seat. This produced initial wave pressures in the gun of a highly dangerous character, and so these results led to the size of the grain being increased, with the result that less burning surface was exposed, but as the pressures still remained high, chambering, or enlarging the powder chamber of the gun, and so giving more air space for the charge, was tried, with somewhat satisfactory results, as the pressure by this means was kept down; but, still, it was not wholly satisfactory, as much unconsumed powder left the muzzle with the projectile. Built up charges of prismatic powder were then introduced, with a hole bored in the centre of the prisms, the idea being that as the prisms burnt from the outer edge and inner surface, that a constant area of burning surface would be exposed for the generation of gases, which would result in all the prisms being consumed as the projectile left the muzzle, and that it would produce a constant and low pressure on the projectile up to the last moment. Black prism powder was the result of this movement, but as the size of guns and charges increased, this gave a much too great initial pressure, and the result of experiments based on alterations in the ingredients of gunpowder produced cocoa powder, so called from its cocoa-like colour. It contained :

Saltpetre	79 parts.
Charcoal	18 „
Sulphur	3 „

The charcoal used was made from straw carbonized by steam, and the carbon so produced was enabled to hold the requisite amount of moisture required.

The service name for this powder is Prism, Brown. As the guns and charges still increased in size, it became necessary to introduce a still slower-burning powder than prism, brown, and certain small alterations in the manufacture of cocoa powder produced slow-burning cocoa (S. B. C.), and E. X. E. (Experimental E.), prism powders (fig. 68).

The S. B. C. prism is of a cocoa colour with an indentation round the core.

The E. X. E. prism is of slate colour with a groove round the core.

In all the powders, in addition to the other requisites, such as high muzzle velocity and low pressure, the following points had to be kept in view; uniformity of action, which is remedied by blending; freedom from fouling and non-liability to damage by transport; ability to retain its moisture when stowed in magazines with high temperatures.

A fair idea having now been conveyed of how these powders are produced, the next step is to discuss, in a general way, how they are adapted to service requirements.

The charges for a gun are invariably made up into cartridges, before being put into a gun, either whole, in halves, quarters, or eighths.

For instance, the full charge for the 111-ton 16·25-inch breech-loading gun is 960 lb. of S. B. C. and this is made up by placing eight cartridges of 120 lb. each in the gun.

In the 45-ton 12-inch breech-loading gun, four cartridges of $73\frac{3}{4}$ lb., prism, brown, are placed in the gun to make up the full charge of 295 lb.

In attacking armour, full charges would be placed in the guns, but if great penetrating power is not required, such as firing shrapnel shells, then "reduced" or three quarters of the full charge is used, which, in the case of the above-mentioned guns, would be six and three cartridges respectively; for general practice from guns during peace time, half charges, or practice charges as they are called, are used.

With the exception of prismatic charges, the amount of powder required is weighed out and inclosed in a cartridge bag, made of silk cloth, which is entirely consumed when the cartridge is fired, otherwise a burning residue would be left, which might cause a premature explosion when the next cartridge was placed in the gun.

The cartridge bags, after being filled, are choked (*i.e.*, the mouth sewn up with silk twist) and the surplus bag cut off to prevent its overlapping the edge of the cartridge and perhaps jamming it in the gun; it is then hooped with silk braid, to preserve its shape, and marked with its nature in printer's ink. When inserted in a gun, the choked part must always be away from the tube that fires it, so as to let the single part of the bag take the flash of the igniting tube and be easily penetrated.

Red shalloon is also used as a powder container for small charges; it is strong and easily penetrated by a flash.

Prismatic charges are built up, by placing the number

of prisms to give the required weight on a false bottom of the required size; a cartridge bag is then drawn over them, the charge inverted, the false bottom removed, and the top of the bag sewn on. When brown prismatic powders are used, as they do not readily ignite, a few black prisms are built in the centre, just opposite the vent, where they are the first to receive the flash, and their ignition is sufficient to ignite the whole charge.

The powder, as has been described in the previous chapter, is stowed in the ships' magazines in powder cases, and there is an exceedingly large variety of these in use in the service.

There are a few general conditions that hold good for all of them; they should be strong, and both air and water-tight. They are made of both wood and metal.

For instance, two $157\frac{1}{2}$ lb. S. B. C. cartridges for the 13·5-inch breech-loading guns, are stored in a T metal cylindrical case.

Four $41\frac{1}{2}$ lb. prism, brown, cartridges for the 9·2-inch breech-loading Mark VI. gun are stowed in an I rectangular case.

Fourteen $7\frac{3}{4}$ lb. S. P. cartridges for the 5-inch breech-loading guns are stowed in a whole metal-lined case.

The cartridges are either pushed out of the cylindrical cases into a cage, which is raised by hydraulic pressnre to the gun, or they are put into Clarkson's leather covered cartridge cases, and taken from the magazines to the guns.

The condition of the powder is checked by taking the pressures set up in the guns whenever full charges are fired. This is done by means of inserting two crusher gauges in the cartridge before it is placed in the gun.

The crusher gauge is a strong steel cylinder closed at one end by a screw plug, in which a gas-tight movable piston works. This piston bears on a small copper cylinder, and is forced in by the pressure set up on discharge of gun, thus crushing up the copper cylinder. The original length of this copper cylinder is half an inch, and it is crushed by hydraulic power corresponding to 9, 12, or 15 tons; a copper just below the pressure expected in the gun is placed in the crusher gauge, and after discharge the steel cylinder is withdrawn from the bore of the gun, and the copper extracted and measured by a micrometer. The amount of compression shows the pressure set up, and is extracted from a table of pressures corresponding to compressions.

The normal pressure in a 13·5 breech-loading gun, using 630 lb. of S. B. C. with a 1,250 lb. projectile, is 17 tons, and the muzzle velocity 2,016 feet-seconds; with a 4-inch breech-loader using 12 lb. of S. P. and a 25 lb. projectile, the pressure is 14 tons, and the muzzle velocity 1,900 feet-seconds.

When quick-firing guns and torpedo boat attacks became a recognized feature of naval warfare, smokeless powders, which until then had been a dream of naval artillerists, became a real necessity, as smoke-producing powders would have nullified the hammering effects of the quick-firers by masking the object aimed at, and would have allowed torpedo boats to creep up in the smoke produced.

The French were the first to adopt a smokeless powder, and for some time the secret of its manufacture was most jealously kept, but as soon as the necessity arose smokeless powders of various sorts soon made their appearance in the market, although it was some considerable time before one could be produced that would meet the many and various requirements of a service propellent; this eventually made its appearance in the shape of cordite, but from the criticisms and litigation that heralded its introduction, its nickname of dis-cordite was at that time more appropriate (fig. 68).

Cordite is the result of experiments carried out by Sir Frederick Abel and Professor Dewar. It consists of

58 per cent. of nitro-glycerine,
37 ,, ,, gun-cotton,
5 ,, ,, vaseline.

Nitro-glycerine is an oily, colourless liquid, and is made by mixing a little less than double the quantity of nitric

CORDITE.

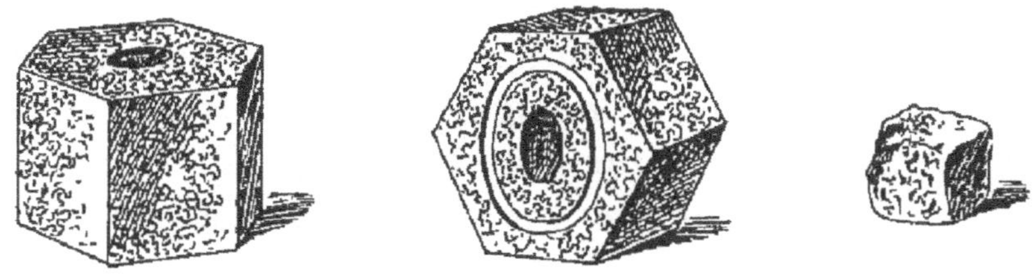

PRISMATIC. E. X. E. PEBBLE.

FIG. 68.—POWDERS.

acid with sulphuric acid, and allowing it to cool; about an eighth of the weight of glycerine is then added gradually, so as to keep the temperature low, and the mixture is kept below 70° Fahr. by passing air and cold water through it. When the mixture has stood long enough the acids are drawn off, and the nitro-glycerine is washed and filtered.

It is an active poison, and produces a violent form of headache if you are brought into contact with it or its vapour. It cannot be easily ignited by a flame, and a lighted match if plunged into it will be extinguished. It is very sensitive to friction and percussion, both of which will detonate it, and the higher the temperature the greater the sensitiveness.

It solidifies at about 40° Fahr., and the greatest care is required to thaw it. Its explosive force is slightly over twelve times that of gunpowder.

The gun-cotton used in the manufacture of cordite is made at Waltham Abbey.

The raw cotton is "teased" or torn into shreds, and cut into 2-inch lengths; it is then dried and dipped into a mixture of sulphuric and nitric acid, from which it is taken and placed in a stream of water. When it has been completely washed, it is wrung out in a centrifugal machine, and again placed in water and washed, being again wrung out. It is then boiled, dried, pulped in a cutting machine, and pressed into discs. When these operations are completed, not the slightest trace of an acid should be present, and this is tested by means of litmus paper, which changes colour on being brought into contact with an acid. Vaseline is obtained from petroleum oil, and its use is to lubricate the bore of the gun and thus lessen the friction set up between it, and the projectile as the latter travels along the bore. It also gives a waterproof nature to cordite, and assists in producing a mixture that is easily squeezed out into a cord.

Acetone is used as a solvent in the manufacture of cordite. It is a colourless liquid, prepared from acetate of lime, and must be free from all impurities.

To manufacture cordite, the requisite proportion of nitro-glycerine is poured over the gun-cotton, and the two are kneaded together, with the addition of acetone, which converts it into cordite paste. To this the vaseline is added, and the whole thoroughly mixed up, after which it is put into a press machine, from which the cordite is pressed out, and either cut into the required lengths, or if of the smaller nature reeled up. It is then dried, and, if necessary, blended by mixing the different natures together.

The nature of cordite is represented by a fraction, the numerator of which gives in hundredths of an inch the diameter of the die through which the cordite is pressed, and the denominator, the length of the sticks in inches.

Thus nature of cordite shown as $\frac{30}{14}$, means that its diameter is thirty hundredths, or three-tenths of an inch, and is cut in 14-inch lengths. This description is used for the 6-inch quick-firing gun.

Many and great difficulties have been met with and overcome in the manufacture of cordite, and it is now unquestionable that the most satisfactory propellent of modern times has been secured for the use of our Navy, and in a few years gunpowder, as it is now understood, will practically have vanished as a service store.

Cordite is manufactured in long sticks or cords, and has the appearance of an opaque and elastic looking substance, the smaller natures closely resembling the uncovered elastic used in spring-sided boots. The diameter of the cordite used increases with the size of the gun it is employed with; thus for a 6-inch gun cords of 0·3-inch diameter, and for a 3-pounder cords of 0·05-inch diameter are used. The charges for the smaller nature of quick-firing guns are made by binding the bunch of cordite until the two ends meet, and the whole can be forced into a cartridge case. The larger nature of charges are made by cutting the cordite into required lengths, and tying them up into a bundle.

It is necessary to use an igniter of powder to cause the explosion of a cordite charge, and so a shalloon bag, containing fine-grain powder, is secured to the cordite in such a way that the flash from the tube firing the gun will ignite it, and cause the explosion of the charge of cordite.

Cordite can be held in the hand and lighted without danger, and is indeed one of the safest explosives known; if ignited by itself in air it will burn without smoke or residue and with a bright flame, and some time is occupied in the consumption of even one inch.

It has been decided that cordite can never be kept in a temperature exceeding 100° Fahr. without deterioration. This at first caused great apprehension, as many magazines in the Navy are often over this; but as the necessity for cooler magazines has arisen, the constructors have devoted more attention to their position and ventilation, with the result that a satisfactory lowering of their temperature has taken place. Some authorities hold that cordite can well be kept in a temperature of 130° Fahr., but the naval authorities prefer the certain and lower one.

Cordite has a peculiar action on the bore of a gun in comparison with gunpowder; it causes a washing away of the bore, whereas the gases of gunpowder eat in and pit or furrow it. This defect in the guns is overcome by placing "augmenting strips" on the driving-bands of a projectile when the bore has become sufficiently worn to admit of their use.

Another peculiar effect caused by cordite is the presence of a flame to the rear of the gun, when the breech is opened quickly after the gun has been fired directly to windward; it is more startling than dangerous.

The following example will show the great gain in weight of charge and muzzle velocity by the use of cordite instead of powder; whilst the pressure is not increased, and is generally lowered:

4·7″ Q. F.	Charge.	M. V.	Pressure.
Cordite .	5 lb. 7 oz.	2,180	15 tons.
Powder .	12 ,,	1,780	16 ,,

The full charge of powder for the 12-inch gun was

295 lb., but the cordite charge is only 167½ lb., and the result of the adoption of cordite is that the number of rounds both for the 12-inch and the 6-inch quick-firing guns in the new battleships has been nearly doubled without increasing the aggregate amount of weight in the magazines. The French smokeless powder consists of nitro-cellulose and tannin, mixed with nitrate of barium and potassium in certain proportions

Cartridge cases for quick-firing guns are usually "solid drawn," that is, they are drawn out, during a great many successive operations, from a single piece of metal called a blank. These cases are costly, but it has been found that they can be used several times, even with cordite, which wears them out more rapidly than powder; thus twelve rounds have been fired with cordite from a 4·7-inch case picked at hazard without any apparent deterioration to the case. For 6-inch guns in the Navy the Admiralty have adopted a built-up case, in which the body is made separate from the head, and when both are complete they are united by being screwed together. The advantage claimed for this method is that there are two thicknesses of metal at the point most susceptible to injury, so that if one thickness failed, the other would still perform obturating functions. Moreover, it is just possible a flaw might exist in the metal of the solid-drawn case, but with the built-up case this could not happen. Quick-firing cartridge cases used for separate loading are closed with a lid.

The following will give an idea of the great strides that gunnery has made since the war with Russia.

Then a 68 lb. projectile was fired with a charge of 18 lb. of powder, giving a muzzle velocity of about 1,100 feet-seconds: now a 100 lb. projectile is fired with a 14¾ lb. of cordite, giving a muzzle velocity of 2,630 feet-seconds.

CHAPTER XIII.

PROJECTILES.

PROJECTILES have been used from the earliest ages; indeed, long before the introduction of cannon and gunpowder. Stone shot were hurled from "balistae" and other engines of war, and a simple knowledge of the practice of their craft taught the ancient artillerymen that their rude engines would throw a round shot further and straighter than one of any other form. Solid stone shot were the earliest productions, but were given up soon after the introduction of cannon and gunpowder, although at first used with them. Gibbon says, "stone balls . . . were found too brittle to resist the force of powder."

In 1400, however, the French are reported to have made shot of cast iron, while about this period several combinations were tried and used, such as stone covered with lead or enveloped in iron, lead balls, bronze balls, etc. At Crecy, 1346, the English used three cannon throwing iron balls ("Le passé et l'avenir d'Artillerie"). At the siege of Cadiz in 1596 cast-iron shells filled "with lead, forming projectiles of great strength and density, were thrown from mortars to a distance of $3\frac{3}{4}$ miles." Lieutenant Brackenbury, R.A., in a paper on ancient cannon in Europe assigns the close of the fourteenth century to the general introduction of iron shot, and this may be taken as being approximately correct.

In "The Great Art of Artillery" by Casimir Simienowicz,

1729, red-hot bars of iron are mentioned as being thrown from the town of Marseilles against the besiegers, but the full value of red-hot shot was shown in the siege of Gibraltar, 1779-1783, when the repulse of the grand attack on that place was attributed mainly to their use.

Case shot, or "canister," were introduced probably soon after solid shot, and were merely an improvement on the early methods of firing from guns stones, old nails, bits of iron, bolts, etc., commonly termed "langridge." It is stated that canister shot were first used in the defence of Constantinople about the middle of the fifteenth century. A projectile, somewhat resembling our present day case shot, was used as early as the sixteenth century. It was called "hagel-kügel" (hail shot), and consisted of a cylindrical leaden box with a fuze and some bursting powder round it, the rest of the cylinder being filled with pieces of iron, bullets, pebbles, etc. And the projectile was put into the gun, fuze next the powder charge.

Grape shot also were nothing more than an improved description of langridge, and were used as early as the fourteenth century ("Le passé et l'avenir d'Artillerie"). They consisted of "canvas cartridges containing small balls," and appear to have been generally used until the introduction of "Caffin's grape" in 1822.

Bar shot and chain shot (fig. 69) were also used in very early times, and consisted of two shot joined together with an iron link or iron chain, chiefly for cutting ship's rigging, and were thrown in some cases from a double-barrelled cannon with one vent for firing the charges simultaneously. A curious type of bar shot was taken in the Indian Mutiny, consisting of two shell connected by sword-blades, obviously for use against the personnel of the English. It is in the museum of the Royal United Service Institution.

The shot used for guns were solid until the early part of this century, when hollow shot were introduced by General Shrapnel for use against ships at short ranges, owing to the greater number of splinters produced by them. They were, however, only the common or naval shell securely plugged with iron. They were withdrawn gradually after 1859 from naval service.

Projectiles, for the purpose of this book, may best be treated under two heads, (A) those for smooth-bore ordnance, some description of which will probably be found of interest, although now quite obsolete, and (B) those for rifled ordnance; and each nature will be further subdivided into (1) shot, and (2) shell.

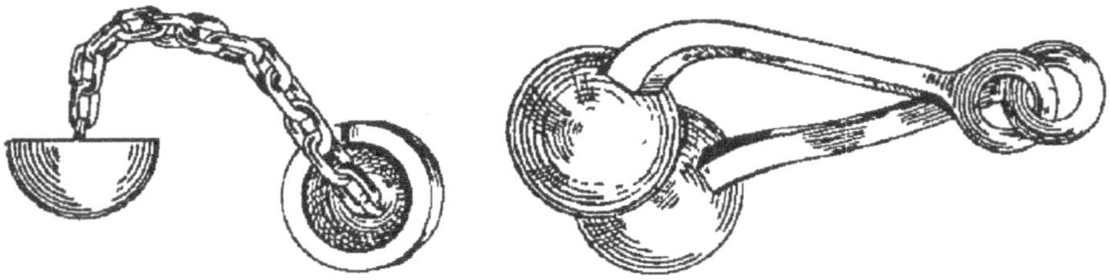

FIG. 69.—CHAIN SHOT. BAR SHOT.

The history of the changes in projectiles may also be said to commence with the present century, and the introduction of rifled guns and the as yet unfinished battle between armour plates and armour-piercing shot has given rise to endless improvements and variations on both sides, chiefly, however, relating to the chemical compositions of the iron or steel used in the manufacture.

SMOOTH-BORE PROJECTILES. (1) SHOT.

Solid shot were spheres of cast iron, and were distinguished by their respective (approximate) weights. With the lighter natures of guns they had wooden bottoms riveted on, and they were used against masses of men, to breach masonry, against wooden shipping, and material generally. When fired hot they were heated to a "wafer" red heat, and fired with reduced charges.

Sand shot were merely small solid spheres of iron of various weights, used chiefly for making up case and grape shot.

Case shot were used for nearly all natures of guns, and consisted of cylinders or cases of iron or tin, with tops and bottoms of wood or iron, the interior being filled with sand shot of different sizes, and the interstices filled with shavings, sawdust, etc. In the case of wood ends, they were nailed on through the cylinder, and the tin or iron ends were secured by hammering over them the fringed edge of the cylinder.

Case shot were used against troops in masses, boats, and to destroy the rigging of ships. They were only effective at short ranges (up to 300 yards or so), owing to the rapid dispersion and lightness of the balls, which soon lost their velocity.

FIG. 70 — CAFFIN'S GRAPE SHOT

Until the introduction of Caffin's pattern in 1822, grape shot consisted of an iron plate, from the centre of which passed up an iron spindle, and round this spindle were piled sand shot inclosed in a canvas bag, which was drawn together by strong line between the balls, and the finished article obtained its name from the rough resemblance it bore to a bunch of grapes.

Caffin's grape (fig. 70) was not, however, manufactured until 1856. It consists of four circular iron plates through which passed an iron spindle screwed to receive a nut at its upper end. These plates had indentations to receive the sand shot, which were piled three in each tier for the smaller natures, four and five for the 56-pounder and 68-pounder respectively, and there were three tiers of shot. The plates kept the sand shot in position, being bolted down and held by the nut on the spindle.

Smooth-bore Projectiles. (2) Shells.

Shells, though not of so early a date as shot, are stated to have been used by the Chinese in the twelfth and thirteenth centuries. They probably were evolved from the "fire-pots" of the ancients, which were mostly filled, it is said, with incendiary compositions. It seems certain that "hand granadoes" were thrown by hand, and by slings, previous to their being used as a projectile from guns and mortars. "Shell" is derived from the German "Schale," *i.e.*, the outside, rind, or bark of anything.

By degrees the use of "granadoes" was extended to mortars, and two sorts were used; the round, retaining the name "granado," and an oblong pattern, being called a "bomb," probably from the noise caused either from the firing of the gun or the bursting of the shell.

Shells, or granadoes, are spoken of as having been used at the siege of St. Boniface in Corsica in 1421, where they are described by Gibbon as "explosive globes," consisting of two hollow hemispheres of stone or bronze, "joined by means of a hinge, a circle of iron, and keys." Voltaire states in his "Essai sur l'Histoire Universelle" that bombs were first used at sea by the French in the bombardment of Algiers (October 28th, 1681).

Various authors state also that they were used in 1521 at the siege of Mézières, 1552 at the siege of Rhodes by the Turks, 1542 at the siege of Bordeaux, and in Holland in

1580: and in 1745, at the siege of Tournay, the French threw shells of 18 inches weighing 550 lb., while the Russians destroyed the Turkish fleet at the battle of Liman on the 17th June, 1788, by their use. A passage in Wilkinson's "Engines of War" seems to prove that mortar shells were not used *at sea* until the close of the seventeenth century.

Various materials have been, also, from time to time, employed for shells—brass, bronze, hardened lead, and even glass, the latter by the Spaniards since the early part of the fifteenth century.

It is worth recording here that shells existed of 19 inches at Boulogne, 1542, of 1,100 lb. weight at Berlin, 1683, of 1,320 lb. at the bombardment of Genoa, 1684. Shells were used at first only from mortars and howitzers, but towards the close of the eighteenth century they were projected from guns also.

There were in the early part of this century several natures of shells, and a short description of each of the principal natures will be given which will suffice to illustrate the type used, although the sizes varied with the guns they were intended to be used from similarly to the shot previously described.

The common shell were hollow concentric spheres of cast iron, with a fuze hole for inserting a fuze to explode the bursting charge contained inside the shell. At first the fuze hole was very roughly screwed, but when metal fuzes became more in vogue, more accurate threads were required, until eventually the gun-metal bushing to the fuze hole became a necessity. Wood bottoms were also used with shells as a means to keep the fuze in its proper position as nearly as possible. These shells were used either as missiles or as mines; as missiles against the personnel of an enemy, as mines against the materiel. Against shipping they were very effective, not only destroying rigging and men, but the smoke from the burst shell in a confined space gave great trouble and annoyance.

Common shell when supplied for naval purposes had the fuze hole always bushed, and the arrangement of the rivet holes for the wood bottom was different, as it was necessary when "double shotting" the guns to have both projectiles touching each other, metal to metal, and the wood bottoms were hollowed out for this purpose. When two projectiles were placed together in the bore of a gun it was said to be "double shotted," but the natures of these projectiles varied thus: two solid shot, a solid shot and a shell, a solid and a grape shot, or a double charge of case or grape, were all combinations used in the Navy. Some naval shell had wood tops instead of bottoms, but the object was similar in all cases, viz., better stowage, and prevention of injury to the fuze. At first the fuze holes were cylindrical in gauge, but with the advent of the Pettman fuze came the taper gauge, and all shell were bushed with a gun-metal adapter, cylindrical outside and tapered inside.

Mortar shells[1] were similar to common shells, excepting that they had a larger fuze hole and no wood bottoms.

Shrapnel shells were invented by General Shrapnel, Royal Artillery, and were at first called spherical case shot. They were the outcome of the siege of Gibraltar, 1781, when it was found that shell fired from guns in preference to howitzers and mortars had a much greater effect owing to their higher velocity, and General Shrapnel saw that if filled with bullets with a bursting charge of powder just sufficient to open the shell a vastly increased effect might be obtained. In 1803 his invention was tested, and successfully used against the French at Vimiera in 1808. In

[1] Mortars were often used afloat, being mounted in mortar-boats and on rafts. Mortar-boats were used during the last war with Russia.

the first patterns the bullets and powder were mingled together in the shell, but as this was found to cause premature bursts owing to the friction set up between powder, bullets, and shell, Colonel Boxer suggested the separation of the bursting charge from the bullets, at first by placing the powder in a cylinder in the continuation of the fuze hole (fig. 71), and afterwards by means of a

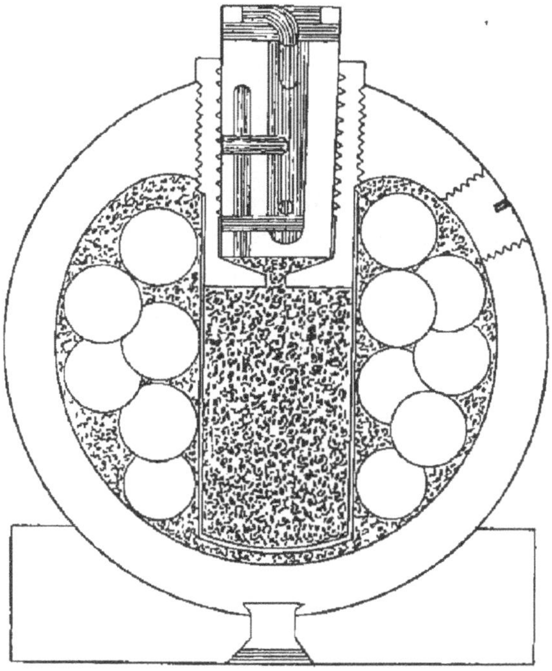

FIG. 71.—BOXER SHRAPNEL, FIRST PATTERNS.

wrought-iron partition or diaphragm (in 1852). The former were called the improved shrapnel shell, and the latter the Boxer diaphragm shrapnel shell (finally approved 1864).

The diaphragm shrapnel shell (fig. 72) was a thin cast-iron shell, divided into two parts, unequally by a thin cup-shaped wrought-iron partition. This partition had a hole in the centre through which a gun-metal socket passed, and the shell was weakened by four grooves down the sides to make lines of least resistance, and so make the shell break up into four or five pieces, and had a thickening at the junction of the diaphragm and the shell to prevent it splitting into only two parts. The smaller partition at the top held the powder, and the bullets were in the larger partition, packed in coal dust. A loading hole for the bursting charge was placed at one side of the shell, and each had also a wood bottom. The use of shrapnel shell was to act as case or grape at longer ranges than were attainable with those projectiles. They have been found a most destructive projectile against troops in masses, and to this present day in rifled guns a modified form of the same shell is used, as will be shown later.

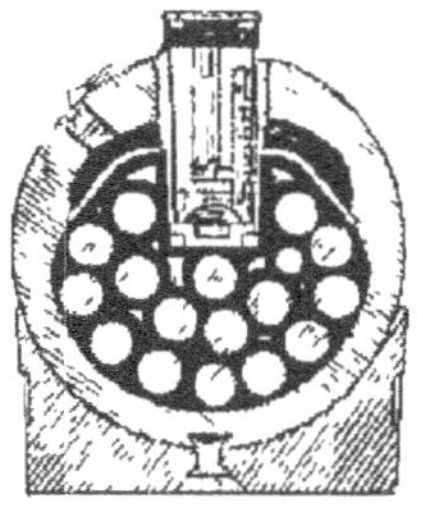

FIG. 72.—DIAPHRAGM SHRAPNEL SHELL.

A number of other missiles were also fired from smoothbore guns, more particularly with the objects of firing the enemy's stores, ships, etc., or lighting up his works, etc. Their origin dates from the most remote ages, when boiling water, boiling oil, Greek fire, and various incendiary compositions were thrown on to ships.

Martin's shell were devised by a Mr. Martin in 1855, as a means of throwing molten iron against an enemy.

It consisted of a spherical cast-iron shell, coated inside with loam, and filled before loading with molten iron, poured in through a filling hole. The thickness at the sides was less than at the top or bottom to facilitate breaking up, and the hole had an enlarged groove to enable the molten metal to seal itself in the shell, and the head was made thicker to make the metal cool, and thus "set" itself in the shell. The concussion on impact broke up the shell, and the molten iron flowed out, setting fire to anything in its path. They were meant to take the place of red-hot shot, and were most effective when filled about five minutes, and were only made for the 10-inch and 8-inch guns. Martin's shell were made obsolete in 1869.

Carcases (fig. 73) were another incendiary projectile, and were spherical shell with three or four vents and filled with a violently burning composition difficult to extinguish. They were made of all sizes. They were first used about the latter half of the seventeenth century. The vents had a sort of fuze composition driven into the shell,

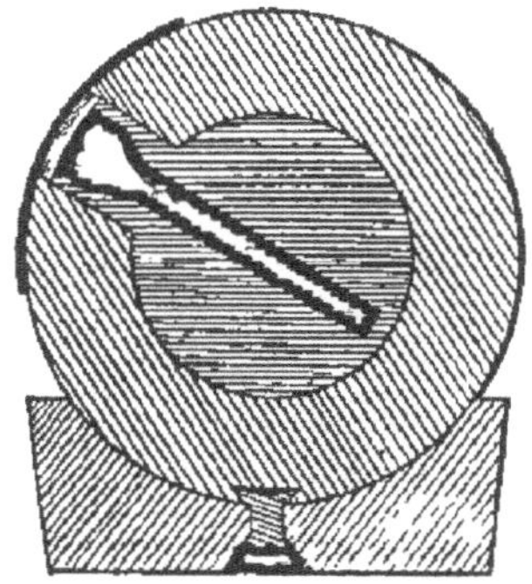

FIG. 73.—CARCASE.

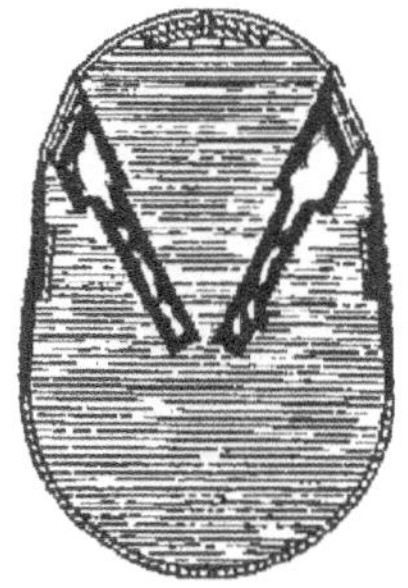

FIG. 74.—GROUND LIGHT BALL.

which was fired on discharge, and gradually ignited the carcase composition inside.

Gronnd light balls (fig. 74) were fired from mortars, and are also of very ancient origin. They were used to light up the ground, and so discover enemy's working parties, and also in place of carcases. They consisted of a wronght-iron skeleton frame, partly covered with canvas, oblong in shape, and the ends formed of iron cups. The interior was filled with an incendiary composition, and had four or five vent holes at the upper end primed with fuze composition. The centre was woolded round, after filling, with twine. They burnt brilliantly when on the gronnd, but were easily extinguished by a few shovelfuls of earth, as were also carcases.

The idea of the parachute light ball or suspended light, originated with Sir William Congreve, and was applied to rockets, but the application in the service form, that of a shell, is due to Colonel Boxer, who in 1850 introduced it to replace the ground light ball, as being impossible to extinguish, owing to its floating in the air. A few were supplied for use during the war with Russia. They consist

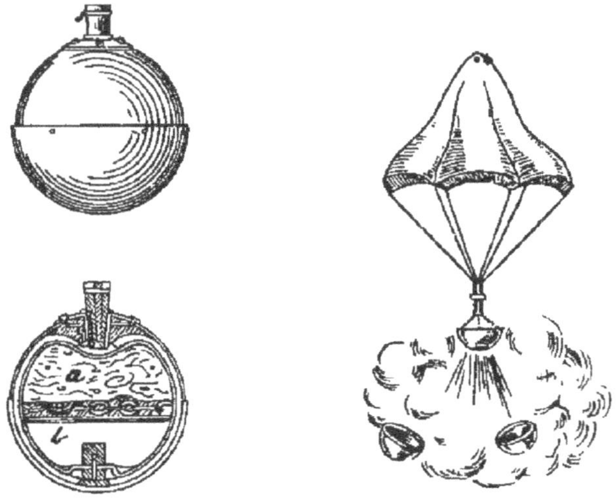

FIG. 75.—PARACHUTE LIGHT BALL.

(fig. 75) of two outer and two inner tinned iron hemispheres, the two outer being lightly riveted together. The two upper hemispheres are connected by a chain, and the inner upper hemisphere has a depression at the top to admit the bursting charge and fuze, and a quick-match leader conducts the flash from the bursting charge to the fuze composition in the inner lower hemisphere. The parachute, tightly folded up, is in the upper part. The fuze being bored so as to burst the shell after it has attained its greatest height of trajectory, and the shell fired from a mortar, on the explosion of the bursting charge, the two outer hemispheres are blown away, and the parachute is opened and expands, carrying the light in the lower hemisphere, which has been at the same time ignited by the quick-match leader.

Smoke balls are of ancient date, as the author of the "Great Art of Artillery," writing in the middle of the seventeenth century says: "We have a way of preparing balls, which, during their combustion, cast forth a noisome smoke, and that in such abundance, that it is impossible to bear it."

In the latter times they consisted of a paper shell, filled with a composition designed to give forth a great amount of smoke, and were used to drive the enemy out of mines, casemates, between decks, and other confined situations. It is, however, doubtful if they have ever been fired at all, and, at all events, they could only have been fired with a very reduced charge.

Manby's shot were projectiles filled with lead, and having a line attached for firing over a ship, and so establishing communication with the shore. They were successfully tried in 1791, and were originally proposed by Lieut. Bell, Royal Artillery, but his claims appear to have been ignored, and in 1811 Captain Manby, R.N., introduced and carried out a similar plan of establishing communication between ship and shore. The shot was at first fired from a small brass howitzer, and was pear-shaped, weighing about 4 lb. 12 oz. The line, which was attached to the shot by raw hide, to prevent the flame of the powder burning it, was coiled so as to run out freely on firing. Later on various sizes were made, with improvements in detail; but in 1865 a rocket, proposed by Colonel Boxer, R.A., was adopted in preference for life-saving purposes, and the Manby's shot soon became obsolete.

West & Co., Southsea, photo. *To face p. 240.*

COMPANY OF BLUEJACKETS AT CUTLASS DRILL. "POINT."

Rifled Projectiles.

The great advantages resulting from the use of elongated projectiles fired out of rifled guns became gradually recognized, and experiments were made to ascertain the best system of imparting a rotary motion to a projectile during its flight. Rifled guns, as has been already stated, first came into use about 1858.

The advantages of rifled over smooth-bore projectiles may be shortly mentioned as follows:

1. A diminished surface for the resistance of the air to act upon, and consequently increased range and greater power at a given range.
2. A flatter trajectory and consequently greater accuracy.
3. The head may be of any required shape.
4. By varying the length, different kinds of projectiles may be brought to the same weight.
5. On the other hand, a specially heavy or light projectile may be fired from the same piece.
6. The capacity for bullets or bursting charge of a shell is increased.

These results were best obtained by the five following systems:

1. A breech-loading system in which the projectile was covered with an envelope of soft metal and forced through the bore on discharge. The rifling consisting of a large number of shallow grooves cut spirally in the gun. This system was introduced by Sir William Armstrong, and is known as the Armstrong or R.B.L. (rifled breech-loading) system, and is only adapted for a uniform twist of rifling.
2. A system proposed by Sir J. Whitworth, in which rotation was secured by the mechanical fit of the projectiles in the bore, which, in his case, was hexagonal, and in the Lancaster gun oval, the gun being bored spirally.

3. A system in which the projectile is fitted with studs or ribs to fit a few deep grooves in the bore. This was used in a B.L. system in France, and in the R.M.L. (rifled muzzle-loading) guns in the service.

4. A system in which the projectile is smooth and is rotated by a copper *sabot* or gas check attached automatically or otherwise to its base. In this case, more and shallower grooves are cut in the gun, and several examples are at present in use in the service.

5. A breech-loading system in which the smooth projectile has a ring or band of copper near its base, which ring projects and is forced through the bore, and taking the grooves, rotates the projectile. The present-day B.L. projectiles are an example of this.

The Whitworth and Lancaster systems have dropped out, and the remaining systems will be referred to in the following descriptions of R.B.L., R.M.L., and B.L. projectiles, these being the three principal types, and each type will be best dealt with separately under its two sub-divisions of shot and shell.

Rifled projectiles can be distinguished by the marks on them. Thus all shells are painted black. When filled, a red band is stencilled near the head. All steel shell have a white band half an inch wide stencilled in addition near the head. Shrapnel shells have a red point painted on. Palliser projectiles have the letter "W" painted on their heads and stamped on the base plugs. Case shot are cylindrical and flat-ended. Armstrong projectiles can easily be known by their lead coating. R.M.L. shells are either studded or studless, with a gas check at the base. B.L. shell will be known by the driving bands of unpainted copper—a yellow band round the centre of a projectile denotes that it is to be used for practice only. Filled shells have the date of filling stencilled on in red near the base. "E" in red denotes emptied and examined shell. All shell having fuzes in them, except 6-pounder and 3-pounder quick-firers, are marked on the head with the word "fuzed."

With B.L. guns after continued firing it has been found that the bore gets worn, and the shell is too small to take the rifling properly. In these cases augmenting strips of copper are used, and are merely, as has been already described, thin rectangular strips of copper hammered into the cannelures of the driving band, and thus increasing the diameter of the shell. They are used with all B.L. and 6-inch quick-firing guns. A special punch is used to make undercut grooves round the bottom of the cannelures for the strips to grip when hammered in.

Early projectiles were usually distinguished by their approximate weight, but as heavy ordnance developed, the nomenclature was changed to the calibre in inches of the gun, still keeping the weight for the lighter natures only. Thus we speak of a 20-pounder R.B.L., 64-pounder R.M.L., 9-inch R.M.L., 4-inch B.L., 32-pounder S.B , etc.

All shells are now lacquered inside to give a smooth surface to the powder, and prevent premature bursting by the friction between it and the walls of the shell. In the larger natures bags are also used to contain the bursting charge for the same purpose.

Rifled Projectiles, R.B.L. (1) Shot.

With the advent of rifled guns came the Armstrong patterns of breech-loading ordnance in 1858. The system of rifling has been already fully described, and gave marvellous results as regards accuracy and steadiness of flight of the projectiles. The twist was necessarily uniform, and the following natures of guns were adopted, viz., 7-inch, 40-pounder, 20-pounder, 12-pounder, 9-pounder, and 6-pounder.

As previously mentioned, the Armstrong R.B.L. system consists of a lead-coated projectile larger than the bore of the gun, and forced through the bore by the lands cutting their way through the envelope.

There are two kinds of shot used with these guns, viz., case shot and solid shot.

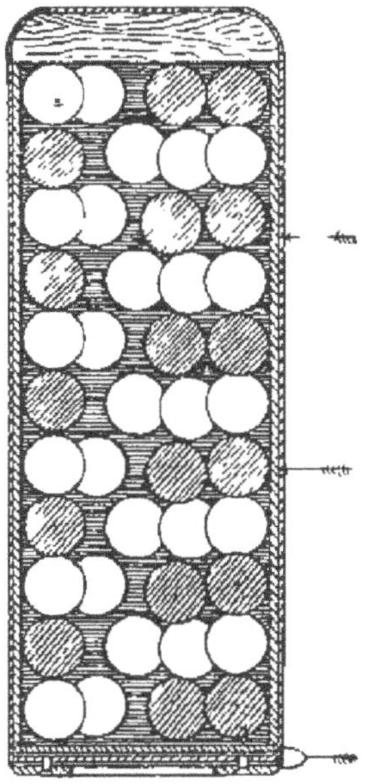

FIG. 76.—ARMSTRONG 12-POUNDER CASE SHOT, MARK IV.

In the case of rifled guns the desideratum is to fire case shot without their taking the rifling of the piece in order to minimize the rotary motion which would tend to spread the bullets of the case shot over too wide an area.

The case shot for the Armstrong guns (fig. 76) consists of an iron cylinder with an iron base and wooden top filled with lead and antimony bullets packed in clay and sand, or coal dust. Solder studs are fixed on the bottom, *not* to take the rifling, but to prevent the shot being rammed home too far down the bore.

The solid shot were used for practice only, and consisted of a solid ogival-headed shot cased in lead.

There are three kinds of shell used, viz., common shell, shrapnel shell, and segment shell.

R. B. L. PROJECTILES. (2) SHELLS.

All the R.B.L. shells have a lead coating outside the shell which is continued over a portion of the base, and a cannelure running round the shell near the base receives any stripping of the lead over the front part.

Various ways were tried for attaching the lead coat to the shell; they were (1) by tin solder and square cut grooves in the shell; this stripped very much: (2) by mechanical means, viz., undercut grooves; and (3) by zinc solder, no grooves. The last was found best as the zinc amalgamates sufficiently with the iron and lead to give a very complete attachment.

These shells were originally made with a fuze hole of the

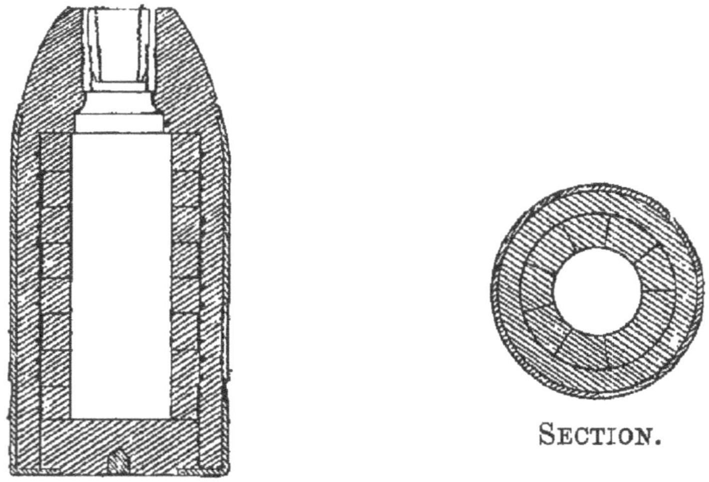

SECTION.

FIG. 77.—ARMSTRONG SEGMENT SHELL.

Moorsom gauge, but are now fitted with adapters, *i.e.*, gun-metal cylinders, screwed outside to Moorsom and inside to general service gauge.

The interior powder chamber was coated with red lacquer. The bursters for the smaller natures of Armstrong segment shell were powder, encased in gas-piping and covered with serge, and were dropped in through the fuze hole.

These shells were used in the wars in China and New Zealand and very favourable reports were made on them. There was, however, in all R. B. L. projectiles, a danger in the lead coat stripping and so making it dangerous to fire over the heads of our own men, but it applied more to the shells with tin attachment than to the later patterns.

The R.B.L. common shell are cylindrical in shape, and the interior is lacquered to protect the bursting charge from premature ignition by friction against the iron.

The manufacture is similar to the segment shell as regards the lead coat.

Common shell are merely locomotive mines made to contain as large a bursting charge as possible consistent with sufficient strength to resist breaking up in the bore of the gun on firing.

Segment shell (fig. 77) were used with all R.B.L. guns and consisted of a very thin cast-iron cylindro-conoidal shell lined with cast-iron segments, built up in layers, and having a cylindrical powder chamber in the centre. The base was closed by a cast-iron disc. The segments were lightly bound together by a thin coating of an alloy of lead and antimony.

R.M.L. Projectiles.

With the introduction of muzzle-loading rifled ordnance which was approved chiefly on the score of simplicity and economy, and owing also to the unsuitability of the R.B.L. system for heavy guns, a different means of rotating the shot became necessary, as it was obviously impossible to ram home a lead-coated projectile. Consequently, studs on the projectile were so placed as to take a few deep and large grooves in the bore. This had the disadvantage of giving a large amount of windage and consequent erosion of the bore of the gun, but has been overcome by the use of the gas check, and, also, this system admitted of an increasing twist being given to the rifling so as to start the projectile more gradually in its rotation and give less strain on the bore of the gun. A disadvantage, however, was, that the studs weakened the shell, but this has been overcome by making the gas check impart the rotation to the projectile instead of the studs, and in the later patterns of R.M.L. shell we find them studless, with an automatic copper gas check which, on discharge, grips the base of the shell and, taking the grooves, so rotates it.

Studs are fixed to the projectiles by "undercutting" the holes for them and then *swedging* in the studs to fill up the holes. In the case of guns having an increasing twist of rifling, the front studs are cut smaller in width than those of the base ring so as to accommodate themselves to the sharper spiral (fig. 78). Two or three rings of studs are given to each studded shell, though as many as five rings of studs were at one time made in shells for the 64-pounder gun. The studs are made of copper or other soft metal.

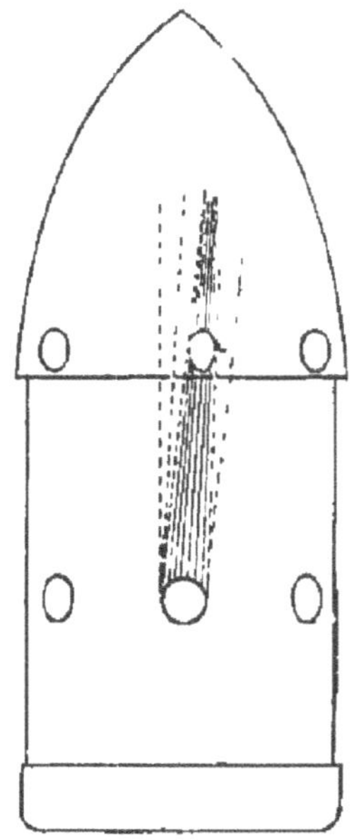

FIG. 78.—ILLUSTRATING STUDS FOR INCREASING TWIST OF RIFLING.

R.M.L. PROJECTILES. (1) SHOT.

Armour-piercing projectiles are made on two principles: first, that of the late Sir W. Palliser, who advocated the use of cast-iron shot, with chilled or hardened heads; secondly, forged steel shot tempered on the Holtzer or Ferminy systems.

Only two natures of shot are used with R.M.L. guns, viz., case shot and Palliser shot, and under the latter heading will be included Palliser shell, which have now been weighted up to and are treated as Palliser shot.

Case shot are used for all natures of R.M.L. ordnance,

and consist generally of an iron cylinder with a base and a head, the interior being filled with bullets packed in clay and sand. The heavier natures have a central stay bolt, and a base-protecting disc. In the lighter patterns the outside lining is made in three segments. The effective ranges vary from 250 to 800 yards.

Palliser shot were designed by Major Palliser in 1863 as a projectile for piercing the armour of ships and forts, and were so successful that they have been, and are still, retained in the service, though in the newer B.L. guns, their place is taken by a steel armour-piercing shot. Palliser projectiles were made in two natures, viz., shot and shell. The latter took a bursting charge, but the difference is only in the size of the cores of the projectiles, which were burst without the use of fuzes. The burster was put into the shell, and a plug screwed into the base filling-hole. When the shell was fired, the bursting charge set back, and was compressed into a solid cake of powder; then if the shell struck a hard substance, the cake of powder flew violently forward, and sufficient heat was thus generated to ignite it and cause the shell to burst. To prevent premature explosion, the core was lacquered, and the burster put in a bag. They consist generally of a cast-iron projectile, with the pointed head cast in a chill mould, so that while the body is of a mottled iron, the head is of a very hard white iron, so hard that no tool work is possible. The base is closed by a cast-iron bush, which is screwed to receive a gun-metal plug, and the studless pattern of shell is rotated by an automatic gas check of copper, which takes the rifling, and being forced by the powder pressure behind on to radial serrated grooves cast in the shell imparts the rotation. Palliser projectiles are not made for guns below the 6-inch calibre. Palliser shot have the advantage of being very cheap, and if they are fired against armour which they can perforate, they break up into innumerable small fragments after having passed through the plate, and sweep everything before them. As armour-piercing projectiles, if compared with the Holtzer or Ferminy, their powers of penetration are but small. For the attack of compound armour, such as is carried by a modern battleship, Palliser projectiles have quite given place to Holtzer or Ferminy. These latter are treated with the sole object of obtaining the highest possible results in penetrating armour plates, and if fired against a ship with old wrought-iron armour of only moderate thickness, they will probably pass right through her, doing but little damage. They are, moreover, very expensive to manufacture.

The earlier pattern of shell had studs, but in the later patterns studs were omitted, and the automatic gas check gave rotation.

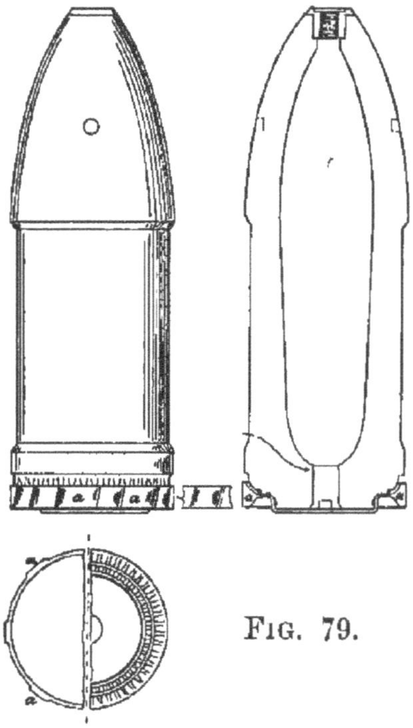

FIG. 79.

COMMON SHELL, STUDLESS, 10-INCH R.M.L. MARK I.

R.M.L. PROJECTILES. (2) SHELL.

Common shells are made for all natures of R.M.L. guns, and are made, studded and studless, of cast iron (fig. 79). The heavier natures have a fuze hole plug at the point, and also a base plug; but the smaller shells have solid bases. The interiors are lacquered to prevent premature explosion by friction of the powder, and, in addition, the heavy natures have the bursting charge contained in a bag, which is first inserted in the shell with its mouth outside, and then filled with the bursting charge. As common shell are required to hold as large a bursting charge as possible, they are made as large inside as is compatible with safety from breaking up in the bore on firing.

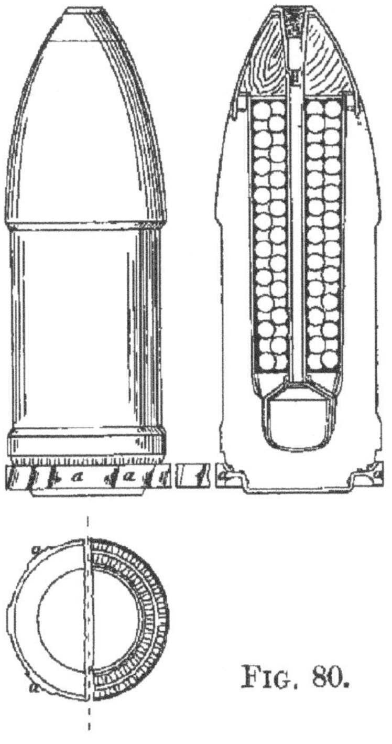

FIG. 80.

SHRAPNEL SHELL. STUDLESS, 10-INCH R.M L. MARK I.

The 7-inch R.M.L. gun has a "double shell," which is merely a common shell of much greater length than usual, and has three longitudinal strengthening ribs inside. It was intended for use against wooden ships, but it is inaccurate in flight at over 2,000 yards. It has two rings of studs, and a bag is used to hold the bursting charge, as in common shell.

The 7-pounder gun also has a double shell, carrying a large bursting charge, which would be effective against houses, etc.

Shrapnel shell (fig. 80) are made for all natures of R.M.L. guns, and are studless, or studded similarly to the common shell. They are exclusively a man-killing projectile. They consist of a cast-iron body strong enough to withstand the shock of firing, with a head of thin Bessemer metal riveted on. This head is filled with wood to preserve its shape, and the point has a gun-metal fuze hole fitted to it. The bottom of the fuze hole has a screwed central hole to take a small metal primer, which serves to convey the flash from the fuze down a length of gas-pipe, to the bursting charge in the base of the shell. The bursting charge is contained in a tin case at the base, and has an iron disc over it to preserve it from being crushed by the weight of the bullets which fill up the shell inside round the gas-pipe. The bullets are packed in melted resin, and a brown paper lining is placed round the interior of the shell to prevent contact between the bullets and the shell. A felt washer covers the bullets at the top.

In the lighter natures of guns the fuze hole socket is shorter, and continued with a tin cone, and a brass tube screwed at the top to take the primer is used instead of the gas-pipe length.

Projectiles for the Modern B.L. Guns.

In the B.L. guns rotation is imparted to the projectile by means of driving bands of copper attached to the projectile near the base. These bands or rings are somewhat larger than the bore of the gun in diameter, and consequently, on firing, the soft metal is compressed into the grooves and cut into by the lands, the copper thus displaced by the rifling overflowing somewhat into cannelures cut for this purpose in the driving band, and at the same time sealing the windage and imparting rotation to the projectile. The projectile has an annular recess at the base, in which are left a number of ribs, into which these driving bands are forced by a press, and, owing to their enlarged diameter, it is obvious that the system is only applicable to B.L. ordnance. A further advantage lies in the fact that projectiles can be cast or forged to gauge, and no further work is required on them, and this allows the hardest material to be employed, as in the case of armour-piercing projectiles.

Two patterns of band have been tried; a narrow band of large diameter, and a broader band of less diameter, with cannelures, already mentioned, to carry off the excess of metal cut from the band by the lands. The latter pattern has now been adopted. The groove carrying these bands is cast in the Palliser and turned in the common and shrapnel shells, and has a number of ribs running round the circumference. The ribs are again divided transversely by being cut away in sections so as to prevent the band from slipping when rotating the shell. After the band has been pressed on, the projectile is placed in a lathe, and the band turned to accurate dimensions. The front edge is tapered, with a view to entrance into the rifling, when the projectile is rammed home. Where guns are loaded by hydraulic machinery at a considerable angle of elevation, as in the case of the 67-ton guns, it is only the jam that takes place between the bore of the gun and the tapered front end of the driving band, which retains the projectile in position after the rammer has been withdrawn.

All B.L. shell which have a base plug have a lead disc hammered over it to fit into a slight undercut recess in the base of the shell, and so completely seal the base hole.

B L. Projectiles. (1) Shot.

Palliser shot are used with all heavy B.L. guns, and are similar in form and manufacture to those for R.M.L. guns (fig. 81). Palliser shell were also made, but the manufacture is discontinued, and those in the service are ordered to be filled up with sand to weight and used as shot. The base plug of cast iron is secured by pouring in molten lead round a ring at the base, and this flows down, filling up an annular recess cut in the base.

Cast steel has been tried for armour-piercing shot, but with little success.

Forged steel projectiles are now made for all B.L. guns of 6 inches calibre and upwards, and will eventually supersede Palliser shot. They have the usual white ring painted on them.

The battle of projectiles *versus* armour, which began with the iron-plated ship and the rifled gun, is likely to continue, and victory is by no means assured to either side as yet, and all the leading artillerists of the great powers are actively engaged in attempting a solution of the problem.

A useful rough rule is Captain Orde Browne's, viz., the extreme penetration to be expected in *wrought iron* is one calibre for every thousand feet velocity per second of the projectile.

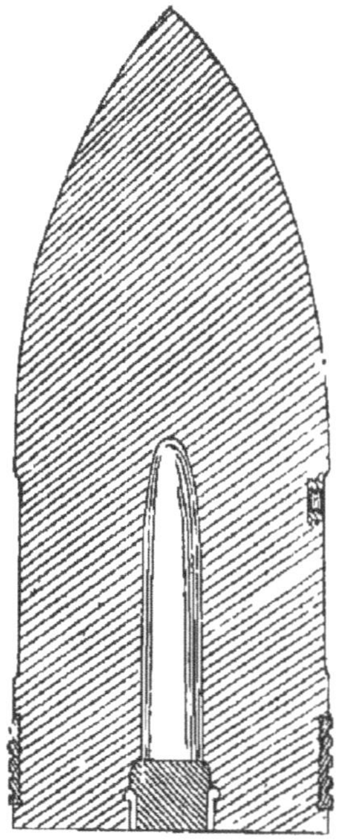

FIG. 81.—PALLISER SHOT, B.L. 12-INCH. SECTION.

The heavy natures of case shot are similar in construction to those of R.M.L. guns, the large natures having a central stay-bolt fastened outside the top by a nut (fig. 82).

Soft metal stops are fixed to the shot near the base to prevent it being rammed too far down the bore. The interior is filled with sand shot packed in clay and sand.

The field service and 4-inch case shot have a copper ring or gas check attached by rivets and studs outside to act as stops in loading.

B.L. case shot are effective at ranges varying from about 1,200 yards to 300 yards according to the size.

B.L. common shell were made of cast iron, but are being rapidly replaced by forged and cast steel shell. Great care has to be taken in designing and constructing common shell to guard against premature explosion. All descriptions of common shell are lacquered inside, and the bursters are put into bags with all natures of guns from 4-inch upwards. The bursting charges used with common shell are generally a mixture of P. and F. G. powders. This mixture gives the highest possible density. The shells, when they burst, break up into large fragments, of which the entire base generally forms one They are supplied by contract, the bushes and driving bands being fitted in the Royal Laboratory. They are similar in design to R.M.L. shell.

The B.L. heavy shrapnel shell are similar also in construction to the R.M.L. types, and the bullets are carried up further into the head. They are made of cast-iron, though steel is taking its place in the 5-inch and smaller calibres (fig. 83). In the smaller natures the bursting charge is carried in a tin cylinder in the head, as in R.M.L. guns, and in the case of steel shell the base is riveted on, and not in one piece with the body. Steel shell do not break up like cast-iron shell, but the body slips off like a glove on the ignition of the bursting charge, leaving the bullets to spread with a slightly lessened velocity. Owing to the absence of the central tube and tin cylinder at the base, a much larger proportion of bullets can be carried than with the early R.M.L. shrapnel.

Ammunition for Quick-Firing.

The smaller natures of quick-firing guns differ from other ordnance in that they take "fixed ammunition," *i.e.*, cartridge and projectile in one metal case. Each nature of quick-firing guns has a shot and a shell and a case shot. Rotation is imparted by a brass ring pressed into a corrugated groove near the base of the shell.

The shot and shell are of forged steel. The shot are for armour piercing, and the shell are made with strong heads for penetrating armour, and in order that their pointed

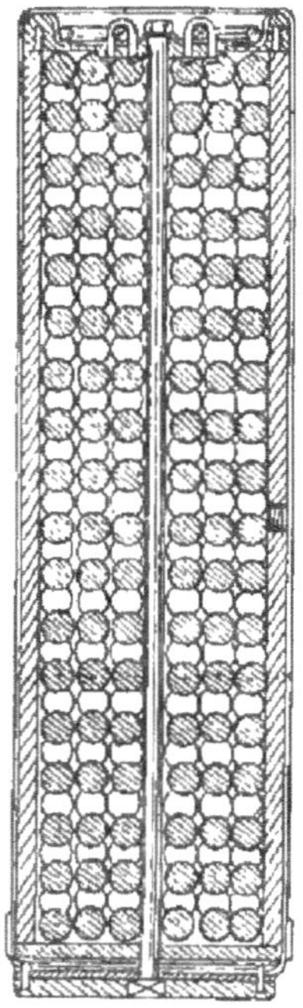

FIG. 82.—CASE SHOT, B.L., 13·5-INCH. SECTION.

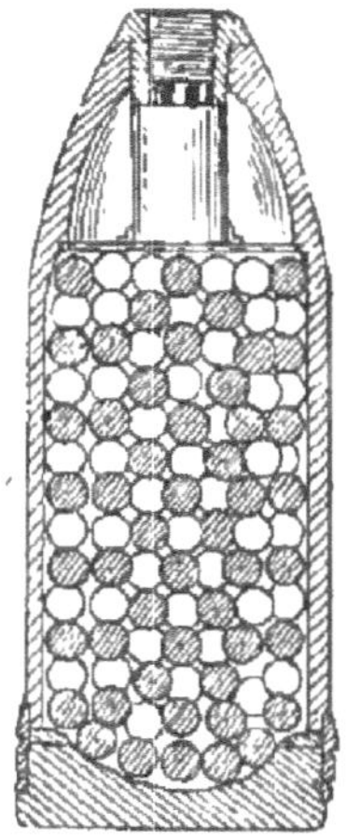

FIG. 83. — SHRAPNEL SHELL, B.L., 5-INCH, MARK IV. SECTION.

heads and strength should be preserved, they are used with base fuzes. They carry as large a bursting charge as the calibre will permit, although of course the capacity of the bursting charge is reduced by the thickness of the head. The case shot is made of brass, lined with three steel segments, and filled with hardened lead bullets packed in clay and sand. Common shell have a flattened point.

Shells Charged with High Explosives.

While gunpowder is being superseded, as the propellent agent in guns, by smokeless nitro-compounds, it also finds a powerful rival in high explosives in connection with bursters for shell. The French here have also taken the lead and adopted melinite, which is a product of picric acid capable of very violent detonation. Up to the present our Admiralty have not issued any high explosive shells to our ships, although experiments have been carried out for some time past at Woolwich with high explosives, which are understood to have given satisfactory results. Melinite shells are practically useless against even thin armour, for the shock of impact causes the shell to detonate before any penetration can take place, and the armour therefore remains uninjured, whereas with gunpowder bursters, ignition is never so rapid, and the shell gets within the ship before it bursts. But for use against the unarmoured sides of cruisers, and the unprotected ends of armour-clads, high explosives are most formidable. Everything in the neighbourhood of the explosion is pulverized, a hole is made in the ship's side of incredible magnitude, and it is quite conceivable that one such shell might decide an action.

Shells with gunpowder bursters break up into larger fragments, and these are more likely to injure material, especially at some distance from the explosion, but it is difficult to estimate the damage that a large shell would do if it burst between the decks of a ship; whether the burster was gunpowder or melinite, it is almost impossible that any living person would escape destruction in the compartment where the explosion took place. Apart from the concussion and flying fragments, the fumes produced by the combustion of the explosive in a confined space would be so poisonous, that, in all probability, those not killed by the fragments would be suffocated, and any combustible material would be set on fire, producing further smoke and fumes. Hence the necessity which has arisen of protecting the secondary batteries of battleships, and the main batteries of cruisers, by armour thick enough to burst "high-explosive" shells before they penetrate. As an instance of the damage which the bursting of a large shell can effect, it may be mentioned that in the battle of the Yalu on the 17th September, 1894, a 12-inch shell from the Chinese ironclad, *Chen-Yuen*, entered the battery of the Japanese flag-ship, the *Matsushima*, hurled the fourth 4·7-inch gun from its mounting, then exploded, firing a heap of ammunition, disabling two more 4·7-inch guns, and killing and wounding ninety officers and men. Several of the other ships were set on fire during the battle by bursting shells, and one result of this battle has been to show the absolute necessity of doing away, as far as possible, with the use of wood in ships of war, a precaution which is now being taken in all the latest types of ships.

Machine Gun Ammunition.

The present service machine guns are the 1-inch Nordenfelt and Gatling, Gardner, Nordenfelt and Maxim guns of ·65-inch, ·45-inch, and ·303-inch calibres.

There are three kinds of projectiles used with these natures of guns, viz., steel, iron, and lead.

The ammunition is "fixed" and consists of a brass case, copper cap, charge, projectile, and envelope of metal.

The projectile is solid, and rotation is imparted to it by the envelope which covers it, excepting the head. The manufacture of iron bullets has been discontinued since 1885.

The ·45-inch and ·303-inch cartridges are similar to those used for the Martini-Henry and magazine rifles.

Fig. 84 shows cartridge for Nordenfelt quick-firing 6-pounder gun (see pp. 163, 255).

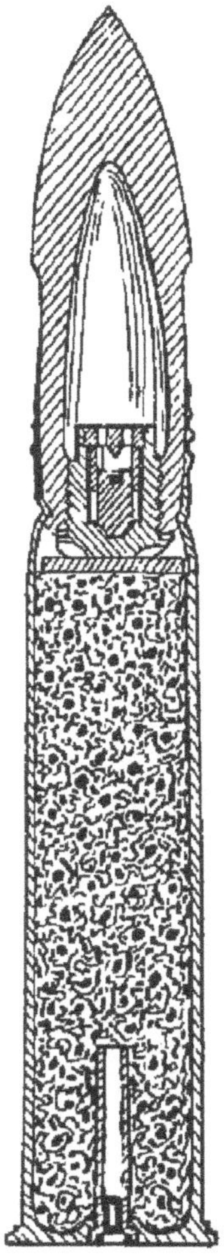

FIG. 84. — CARTRIDGE, NORDENFELT QUICK-FIRING 6-POUNDER.

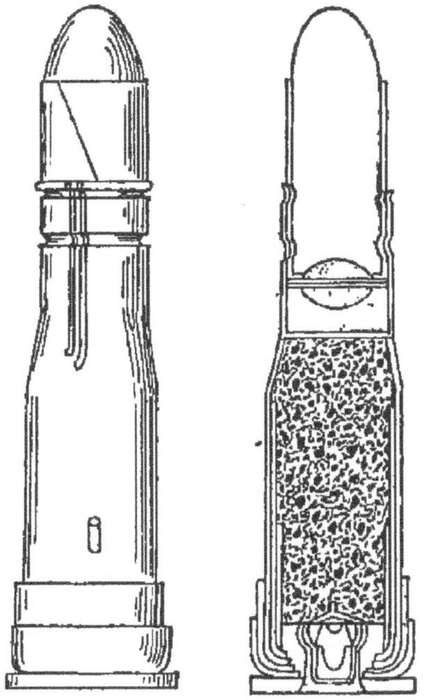

FIG. 85.—CARTRIDGE, S. A. BALL, MARTINI - HENRY RIFLE. MARK III.

SMALL ARM AMMUNITION.

After the days of "brown Bess" came the ·577-inch bore Snider arm, whose cartridge case was made of sheet brass covered with brown paper. The Snider bullet had a hollow head with the lead spun over it, so as to give a sufficient length without increasing the weight, and also to get the centre of gravity in the right place. A hollow was also left in the base to ensure the expansion of the bullet to take the rifling. This ammunition may still be met with in remote parts of the world

The Martini-Henry ·45-inch cartridge at first consisted of a built-up case of thin brass, but has been superseded by the solid drawn case. The bullets are of lead, the early

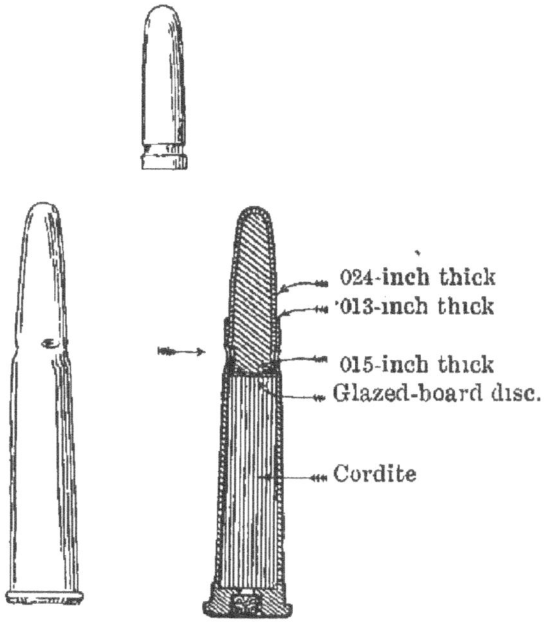

FIG. 86.—CARTRIDGE, S. A. BALL, MAGAZINE RIFLE. CORDITE, SOLID CASE, ·303.

patterns having two cannelures, and the later, one near the base. This cartridge was bottle-shaped. A beeswax wad was interposed between cartridge and powder. This ammunition is still in use and was issued for the M.H. carbine in the Ashanti expedition of 1895 (fig. 85).

Buckshot cartridges are issued with the above calibres for convict guard and other special services.

The ·303-inch magazine rifle has also a solid drawn brass

case, and the bullet of lead has a nickel envelope to give the rotation and to keep the shape of the lead. Two kinds of powder are in use with the ·303 rifle cartridge, viz., black powder and cordite. The black powder is pressed into a pellet with a central fire hole running through it axially. In the cordite charge the space behind the bullet is filled with finely cut strands of cordite. The bullet weighs 215 grains and has a groove near the base. It is held in its place by three indents made in the case, which fit into the groove of the bullet (fig. 86).

Pistol ammunition follows generally the lines of that for the rifle of the time. The later patterns have been for Colt's pistol, and Adams, Enfield, and Webley revolvers. The Colt's pistol ammunition was contained in a skin envelope choked on to the bullet. The remaining natures have either coiled or solid drawn brass cases, and are central fire.

Rockets.

Rockets are still and have been used from very early dates. They are stated to have been used by the Chinese in the thirteenth century, and in Europe in the fifteenth century. To Sir W. Congreve is due the credit of having brought them to their present state of perfection on the system which occurred to him in 1805. Commencing with the idea of an incendiary projectile, he expanded the system to embrace shells and shrapnel for the dispersion of bodies of men and the enfilading of trenches, etc.

On October 8th, 1806, rockets were used against Boulogne, setting fire to the town in many places. Also they were successfully used at Copenhagen, 1807, at Walcheren, Leipsic, 1813, on the River Plate, and against the French boats on the Adour. The action is due to the difference of pressure between head and base, or as the result of the pressure of the gas generated against the air "as a fulcrum." The difficulty of keeping this missile point first was first overcome by attaching a stick to the side or base by which the centre of gravity is brought so near the point, that pressure of the air acts on the stick with sufficient leverage to keep the point always towards the direction from which such pressure proceeds. The second method was by imparting a rotation to the rocket, and will be described later.

The Congreve rockets were made first of paper, then of iron, and were named according to their weights.

On September 14th, 1864, the Congreve rockets were superseded by a rocket of General Boxer's construction, with certain improvements in the vent disc at the base, and a stronger composition.

These were again superseded, in 1867, by Hale's rockets, 9-pounders and 24-pounders, and a description of these is here given (fig. 87).

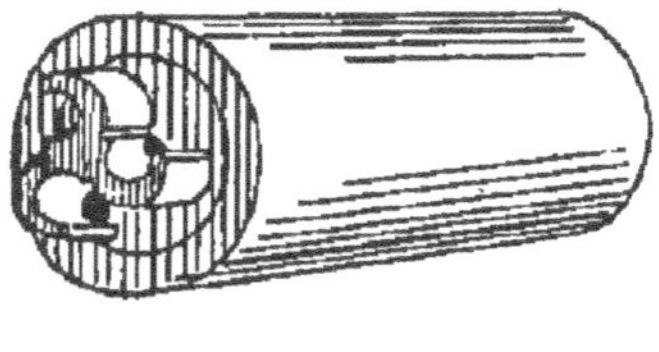

End View, Showing Tailpiece.

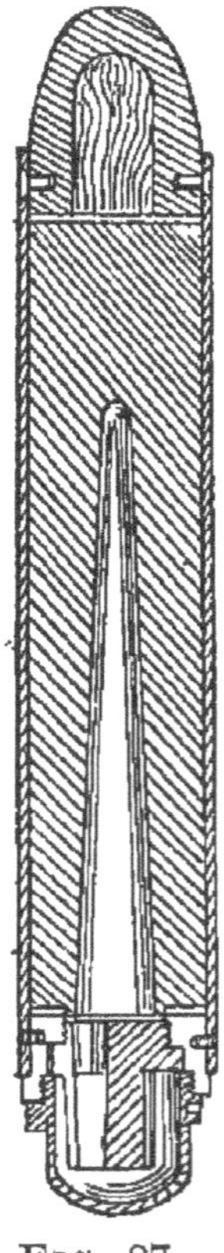

Fig. 87.—24-pounder War Rocket. Mark VII.

The head is of cast iron, plugged with wood, and riveted on to the body of atlas metal (a mild steel). The interior is lined with brown paper and calico to prevent contact of the metal and composition. The base is closed by an iron disc, and the disc is tapped to take the tailpiece. The tailpiece is of cast iron, cupped out inside, and contains three conical vents, the larger part of the cone being towards the interior of the rocket. The vents are cut away on one side, and in consequence of this, the gas meeting no resistance where the cone is cut away, gives rotation to the rocket and so steadies it in its flight. The body is filled with composition, driven in by hydraulic power, and a conical hole is left in the latter to give a large surface for ignition. Early patterns had three corrugations on the case to better hold the composition.

Rockets are very useful for setting places on fire, and are especially valuable against savage tribes.

They are fired by friction tube from a trough on land or from a tube in boats.

The 12-pounder Life-Saving Rocket was introduced by General Boxer in 1865, and consists of two rocket bodies in prolongation to give great length of burning withont any sndden violence, which might break the line carried. The rocket carries a stick and the line is attached to the top end of it, and passes through a hole at the bottom. There is no tailpiece. The line is carried in a box, coiled on pins so that it can run off freely.

CHAPTER XIV.

FUZES.

Fuzes are used with hollow shell as a means of igniting the bursting charge of gunpowder. The present-day fuzes, which are screwed into either the head or the base of the shell, contain their own means of ignition, but formerly they were ignited by the flame of the cartridge which passed over the shell on firing, and still earlier we find instances where the fuze in the shell was first lighted by the gun detachment before the cannon was fired.

One of the earliest forms of fuze is mentioned in Gibbon's "Artillerist's Manual," where an iron fuze is said to have been used at the siege of St. Boniface, Corsica, in 1421. It is spoken of as "a sheet-iron tube inclosing the priming, and riveted on to one of the hemispheres of the shell."

In Machiavelli's "The Art of Warre," published first in Italy, and translated by Peter Whitehorne, with addenda of his own and dated 1560, is found the following "Maner how to make trunkes of fyre," and describes a sort of wooden shell, made of cane, as big as a man's thigh, fastened to the staff of a partizan and filled with alternate charges of serpentine powder and rosin, camphor, and broken glass, covered at the head with parchment with a small hole "wherein you must put a matche made with gunpoulder, the wiche easely and quickely with your common matche maye kendell the fyre."

And again, speaking of hand-grenades: "Certaine hollow balles of mettel as bigge as smalle boules, and a quarter of an ynche thicke, caste in mouldes and made of three partes of brasse and one of tinne . . . whiche balles filled half full of fyne corne poulder, and the other halfe ful of serpentyne poulder, myngled with rosen beaten into poulder . . . and then puttinge in the mouthes of the holes of them a littel fyne corne poulder to make the rest to fyre the suner;" says they will break and fly into a thousand pieces, and also recommends that they are thrown quickly when lighted, lest they hurt those throwing them.

The fuzes of these primitive shells were the "matche" in the first and the "fyne corne poulder" in the second instance.

In "The Compleat Gunner," 1672, by Captain Venn, mortars are mentioned as being useful to throw "Great Granadoes" of 100, 150, and 170 lb. weight, and also smaller of 40 and 50 lb. A print in this book also shows a mortar shell with a fuze in the hole, and the author speaks thus of hand granadoes fuzes: "The largeness of the orifice [scil. of the shell] in which you must put in your fuze made of wood whose upper part must be about $\frac{2}{9}$ the diameter of the Granado, and the small hole in the Fuze should have the largeness of $\frac{1}{18}$ of the same diameter. The rest of the capacity of the Shell must be filled with well grain'd powder. The length of the Fuze must be $\frac{2}{3}$ of the diameter and the top must be broad and a little rounding like a Hemisphere. The Hollow and inner part of the Fuze must be about $\frac{1}{9}$ diameter at the small or inner end and $\frac{2}{9}$ at the outer end."

In "A Light to the Art of Gunnery," 1689, by Captain Thomas Binning, we find a description of loading mortars and placing turf or rope-yarn between the shell and the charge so that the light from the fuzee may not ignite the powder, and also, "when you would discharge a mortar peece—first you must set fire to the feusee in the Granado [scil. shell] and you must see it burn well before you give fire at the touch-hole."

However, in the "Memoires d'Artillerie," by Surirey de Saint Remy, 1707, p. 309, we find a description of a fuze translated thus: "Fuzes for Bombs of 12 inches diameter are of lime, willow, or alder, well dried and without splits. Although in these sorts of woods are many small knots and cracks which make them defective, these woods have other properties which necessitate their use. These fuzes must be well and cleanly bored, and shaped outside and inside, for it is commonly found in fire channels, when they are not well bored by a workman who has special tools, that fragments remain which are very detrimental, for when charging the fuzes they mix with the composition, rendering it subject to extinction, and when found they must be pushed out with a stick. Bomb fuzes are made of two lengths $8\frac{1}{2}$ inches and $9\frac{1}{2}$ inches, at the small end 14 lignes thick (a ligne is $\frac{1}{12}$ part of an inch) and at the large end $18\frac{1}{2}$ lignes thick. The fire channels of both are 5 lignes diameter."

The powder was reduced to dust and rammed tightly into the fuze, and a covering was used to protect the head.

In a book written in French, 1839, published at Brussels, called "Memorial d'Artillerie," appears the following: "Sometimes the Fuze comes away from the Shell. . . . When firing 7-pounder shell under 15 or 10 degrees elevation from $\frac{4}{10}$ to $\frac{5}{10}$ of the fuzes do not entirely burn, and at from 2 to 5 degrees of elevation about $\frac{1}{8}$ of the fuzes do not entirely burn;" *i.e.*, the shells do not burst. This is stated to be the opinion formed by the writer, based on what he saw of gun practice at Breslau in 1812.

In the "Traité d'Elementaire d'Artillerie, by E. Decker, 1825, occurs a passage translated as follows: "The fuzes used in France with the 'obus' or hollow bullet, are made of cones of light wood with the head cupped out and turned, and a hollow axis running down not quite the whole length of the fuze. This centre channel is filled with a composition, and is cut to any required length." It was made in three sizes for the French artillery.

In "Heavy Ordnance," published 1837, by Captain T. F. Simmons, R.A., occurs the following: "To the necessary precaution of a box for each shell, the French introduced into their service, in 1824, a cap with a screw for the top of the fuze, which is only taken off when the shell is actually placed in the muzzle of the gun; this fuze has been adopted into our service. . . ." And further on, while advocating the employment of fuzed shell in our Navy, he writes: "If shells, even in very limited numbers are introduced into our Navy . . ."; showing that our Navy had not then adopted fuzed shells.

Fuzes may be best discussed under three heads. (1) Time fuzes; (2) percussion, or concussion fuzes; and (3) time and percussion combined fuzes, and we will enumerate them in this order.

As during the present century nearly every ardent artillerist has tried to earn distinction by inventing either a fuze or a range-finding instrument, the number of ideas and suggestions which have been tried or rejected is legion. It must suffice, therefore, to outline the features of those which have had at any rate a partial success, and have been used in the service, and one fuze of each type will be described.

(1) Time Fuzes.

We have seen that the earliest forms of time fuze consisted mainly of a cone of wood or metal, open at the top, and with a centre channel of composition, which was cut or sawn off to any required length, roughly guessed at. And this form of fuze, while tardily adopted in the English service, seems to have been the only type in existence until between 1829 and 1832, when some improved gun-metal fuzes by General Miller were adopted for all filled naval shells.

These fuzes were cut or sawn off (a dangerous proceeding in the case of metal fuzes) to the required length, and, in the case of wooden fuzes, the exterior was either rasped down or wrapped round with tow to fit the fuze-hole of the shell, and was then hammered in tightly. The metal fuzes were always screwed in, and were chiefly those used by the Navy. The shell was then placed in the bore, and the cap, if any, removed, and the shell rammed home. Numbers 3 and 4 at the gun were responsible for the cap being removed before ramming home. Three natures of naval fuze were approved, called the 3-inch, 4-inch, and short-range fuze, whose total times of burning were 7 seconds, 20 seconds, and 2 seconds respectively.

In 1852 the metal fuzes had side holes leading to the central channel which were stopped with clay or putty, and which, when cleared out, allowed the flame to pass into the shell through whichever hole was cleared. With these rough and ready methods of fuzing and fixing the fuzes it is not surprising that nearly every other shell burst at the muzzle.

The times of burning of given lengths of composition were arranged for by varying the ingredients of the composition.

Wood was principally used for land service and metal for naval shells.

Mortar Fuzes.—The mortar fuzes were simply an improved form of the first type of fuze, and consisted of a cone of wood, with a central channel of fuze composition covered with a cap of tin. Quick-match priming insnres the ignition of the composition, and a spiral row of indentations, numbered, showed where the fuze was to be bored through instead of being cut off as formerly. The action is similar to that described for the Boxer fuzes, detailed further on, except that there were no side channels, and the flame passed directly into the shell from the bored side hole. Two natures were used, the large for the 8-inch, 10-inch, and 13-inch mortars, and the small for 24 and 12-pounder common shell and the Coehorn mortars. They are painted drab colour (fig. 88).

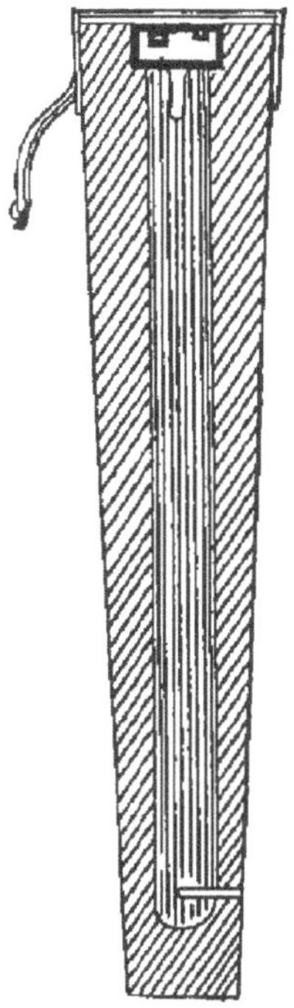

FIG. 88.—LARGE MORTAR FUZE. SECTION.

Boxer Fuzes.—In 1855 General Boxer, R.A., introduced his wood time fuzes, which are in use to this present day, though no more will be made when the stock is used up. They consist generally of a wood cone, with a centre channel filled with mealed powder or fuze composition, and side channels, two at first, but afterwards six and eight in number, filled with pistol powder. These latter channels had radial holes bored at close intervals down the length of the fuze, and the last hole was bored through to the centre channel to insure the fuze eventually bursting the shell. These side holes were stopped with clay, and the exterior of the body of the fuze was covered over with varnished paper, and marked with a spiral ring with dots where the holes were, and numbered, so that when bored through at a given hole, a certain known length of the fuze composition was connected with the side channel bored. The fuze composition was mixed so as to burn 1 inch in 5 seconds, and the mealed powder to burn 1 inch in $2\frac{1}{2}$ seconds. By this method of side channels General Boxer was able to graduate his fuze finer in proportion to the number of channels. The head carried a quick-match or gun-cotton priming, which led the flame passing over the top of the shell on discharge to the fuze composition, and this burned down until the bored through radial hole was met, when it flashed through it and down the side channel, so exploding the shell.

The radial holes were arranged so as to give quarter second intervals of burning, and several natures were made, called after their length of burning the 5, 9, 15, 20, and 30 seconds fuzes, though all are not exactly similar in construction (fig. 89).

It was fonnd that owing to the shrinking of the wood in hot climates a space was sometimes left between the wood and the fuze composition, consequently a compressed paper lining was introduced to prevent the too rapid burning of the fuze. A gun-metal plug was also screwed in to keep the shape of the fuze, and a pin projecting downwards from this held the quick-match priming, which was then led out by two fire-holes at the side near the head and outside the shell.

A short hole was also bored in the head of the fuze composition to insure the flame igniting it by giving a larger surface.

The later patterns have a copper band covered with varnished paper and a tape to protect the head from injury aud ignition before use. These fuzes are about two inches long. They are bored by being placed in a gun-metal borer, in which is screwed down a bit or gimlet, and when screwed home to its stop a hole is bored through to the centre channel. This borer is technically termed a "hook borer," being shaped like a hook, and the bits are replaced when broken by being withdrawn through the handle and fresh inserted.

These fuzes were found to act as percussion fuzes on impact of the shell with any solid body, such as wood, stone, etc.

They were distinguished by their colouring outside; the 5 seconds being painted red and drab, and the 9 seconds

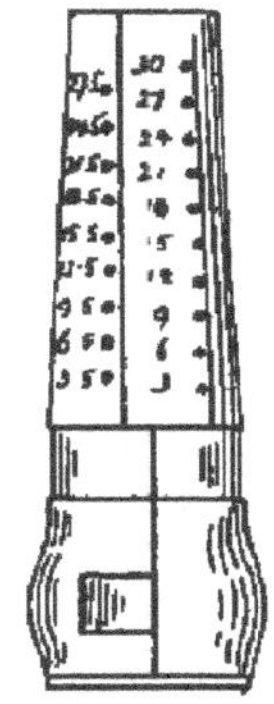

FIG. 89.—15-SECOND WOOD TIME FUZE.

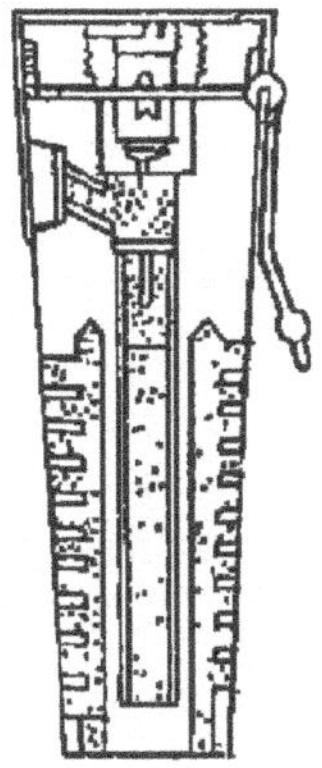

FIG. 90.—15-SECOND WOOD TIME FUZE, WITH DETONATOR. SECTION.

black and drab, etc., and also by the numberings outside on the spiral ring.

Boxer Fuzes with Detonator.—When the Armstrong breech-loading guns came into vogue, owing to there being no windage, a detonator was added to this type of fuze to ignite the fuze composition (fig. 90). A cylinder of alloy was screwed into the head of the fuze; this cylinder contained a hammer supported by a copper suspending wire, and below the hammer was a hollow containing a detonating composition of chlorate of potash, fulminate of mercury, and sulphide of antimony. A hole was bored through the bottom of this cylinder for the passage of the flash, and a copper safety-pin passed through the head of the fuze between the hammer and the detonating composition, so that the fuze could not be accidentally fired even if the suspending wire were accidentally sheared. The pin was withdrawn on placing the shell in the gun. On firing the gun, the shell, moving forward violently, set back the hammer, which sheared the suspending wire and fell on the detonating composition, thus exploding it and igniting the fuze.

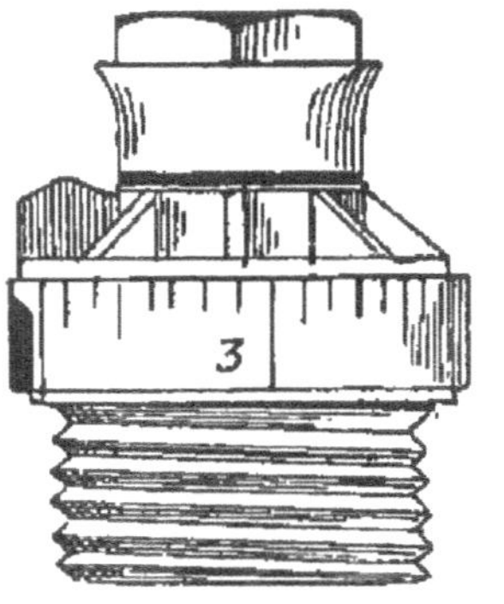

FIG. 91.—ARMSTRONG E. TIME FUZE.

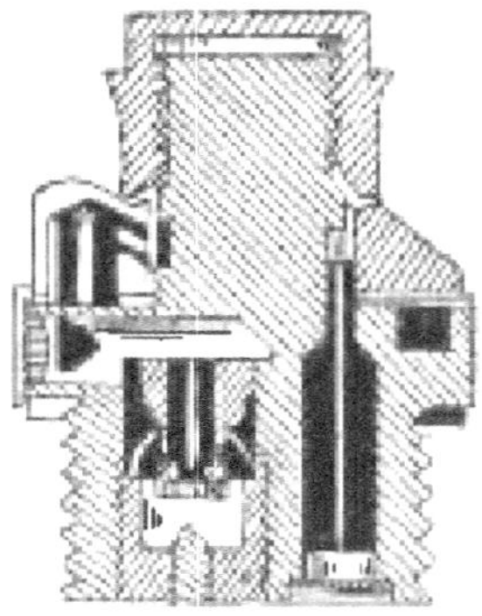

SECTION.

Armstrong E. Fuzes.—With the introduction of the Armstrong guns came also the Armstrong E. time fuze, which was in effect a development of the foreign systems of fuzes, and is the basis on which some of our modern fuzes are made. That for the Navy is here described (fig. 91).

The body and nut are made of gun-metal, and the graduations are marked on the outer rim, and represent inches and tenths. A pellet, supported by a brass cup, is filled with R. F. G. powder, and carries in its head a patch of detonating composition.

The channel by which the flash from the pellet reaches the ring of fuze composition has a strand of quick-match in it to act as a leader. The fuze composition is pressed into a ring or groove running round close to the exterior of the fuze body. This composition burns at the rate of one inch in two seconds, and, owing to a metal stop, can only burn in one direction, from left to right. A leather washer and movable gun-metal collar cover the ring of composition, and at one part of the collar a channel primed with mealed powder communicates with a groove round the neck of the fuze, which groove also contains mealed powder and communicates with the blowing chamber filled with mealed powder and leading into the shell. The movable collar is clamped by a nut on the neck of the fuze. An arrow head outside the collar reads with the graduations round the body of the fuze.

The fuze being set and clamped as required, on firing, the brass cup is crushed in, and the pellet strikes the needle, which explodes the detonating composition, and the flame ignites the fuze composition ring, which burns round till it comes to the channel leading to the groove round the neck, when it ignites the mealed powder, and the flash is instantaneously conveyed into the blowing chamber, and thence into the shell.

This fuze was largely used by the Navy for 20, 12, and 9-pounder Armstrong guns.

Sensitive Fuzes.—The middle and long sensitive time fuzes are the latest development of the metal time fuzes, and are designed to meet the requirements of the breech-loading ordnance, where the shock of discharge is very slight, owing to the steadiness of flight of some low velocity shells entailing a dangerously thin shearing wire. The middle fuze is here described (fig. 92). It consists of a gun-metal body screwed to G. S. gauge with a removable ring of composition running round the top, and held down by a metal dome and screw cap. Immediately below the ring is the igniting arrangement, consisting of a pellet placed in a hammer and retained by the projections on two bolts, which fit into a slot in the hammer, and kept up to their work by spiral springs.

Opposite the detonating patch of the hammer is a steel needle in the rim of the fuze, near the commencement of the composition ring. The bolts are at right angles to the

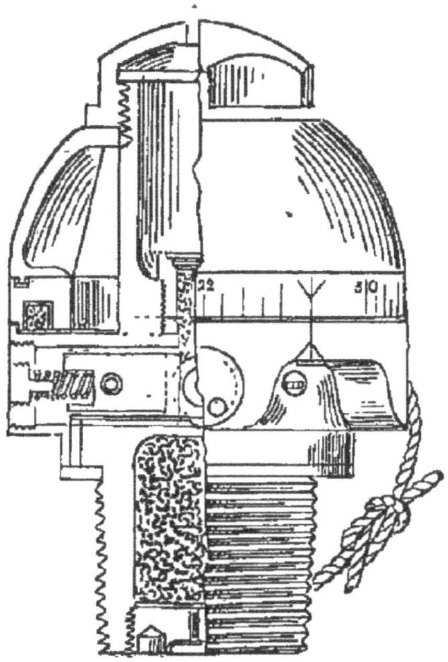

SECTION.

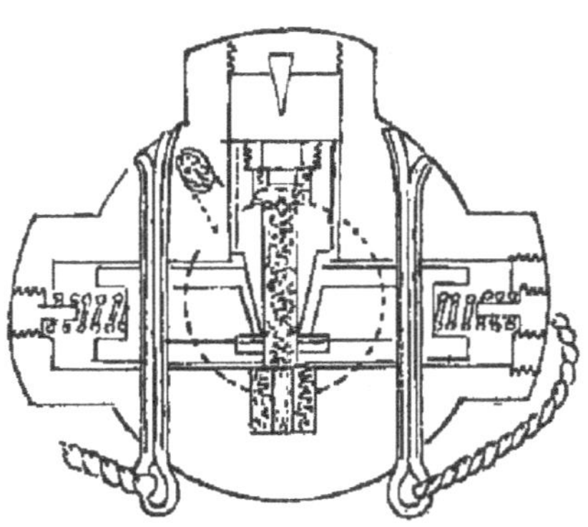

CROSS SECTION.

FIG. 92.—SENSITIVE TIME FUZE, "MIDDLE." MARK I.

hammer, and free to move in slots radial to the composition ring. Each bolt has a safety pin passing through it, both of which must be withdrawn on loading. On shock of discharge the parts are undisturbed, but as soon as the shell begins to rotate the centrifugal force of the retaining bolts compresses the springs and releases the hammer, which, acted on by the same force, flies against the needle, and fires the detonating patch, which ignites the ring of fuze composition which burns round until it meets a hole in the body of the fuze communicating with the chamber in the base, exploding the powder there and passing into the shell through the hole in the bottom plug. This fuze burns up to fifteen seconds, and the long fuze thirty seconds; the construction and action being the same in both cases.

(2) Percussion Fuzes.

This nature of fuze was used as early as the first half of the seventeenth century, *vide* "Great Art of Artillery," written about that period. That for hand grenades consisted of a fuze body, with a number of side holes bored in it Down the centre channel was placed a lead bullet, tied to a length of quick match coiled above it and lighted. On impact the bullet fell through into the shell, and dragged down the quick match, which fired the powder in the shell, through one of the side holes, which were filled with mealed powder.

Another more complicated fuze was used for guns. In the shell was fastened a roughened tube of iron, and in this fitted a plunger or rod projecting from the mouth of the shell, and fitted with two flints, "like the cock of a gun lock." These flints were forced against the sides of the roughened tube on impact, the plunger being pushed in, and the sparks thus fired the shell powder.

Freeburn's Fuze.—The earliest form of fuze actually known to exist in our service was Freeburn's concussion fuze, introduced in 1846 for land service, and was made of wood. The fuze contains a central channel, the lower portion of which was driven with fuze composition, the upper with mealed powder. Three holes, closed with three metal conical plugs, with the large part of the cone towards the interior of the fuze, led into the portion of the fuze driven with mealed powder. On firing, the conical plugs are supported by the mealed powder; this burns rapidly away, and on striking an object the plugs, being now unsupported, fall into the fuze, the flame from the fuze composition passes through the holes and explodes the shell. This fuze was also tried in metal for the naval service. Wood was found, however, to be an unsuitable material for

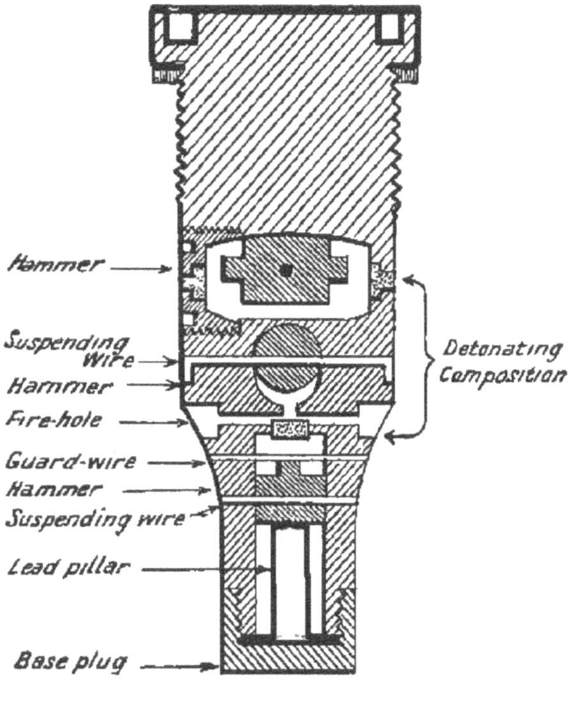

FIG. 93.—MOORSOM PERCUSSION FUZE. SECTION.

a fuze which depends for its action on mechanical accuracy of fitting, and all percussion fuzes were then made of metal.

Moorsom Fuze.—In 1850 the first satisfactory metal percussion fuze for the Navy was invented by Commander Moorsom, R.N., and his principle of wire suspension is still in vogue, although his fuze was obsolete in 1867 (fig. 93).

It consisted of a gun-metal body, cylindrical, and screwed to fit the Moorsom gauge, with a projecting head, and tapering towards the bottom, and about four inches long. Two oval holes were bored transversely, and at right angles to each other, in the body of the fuze below the screwed exterior, and in them were suspended two cylindrical gun-metal hammers with small projections or nipples, which were close to patches of detonating composition placed at either end. These hammers were suspended at the front end of the oval holes, so that the shock of discharge shears the wires and allows the hammers to become free.

At the base of the fuze is a long vertical hole, cylindrical, recessed at the upper end to hold a patch of detonating composition, and a fire hole is made horizontally through the fuze immediately above this recess. A small cylindrical hammer of gun-metal is suspended below this recess by a copper wire, and is supported by a lead pillar at the base A guard wire, also of fine copper, runs across the top of this hammer.

The action is as follows:—The shock of discharge suspends the shearing wires, and the hammers are then free to act. On the shell striking an object the hammers are thrown violently forward, one of their projections being thus necessarily brought into violent contact with one of the patches, which, being exploded, flashes through its fire-hole into the shell, thus firing the bursting charge.

The object of the lead pillar was to act as a sort of cushion to the vertical hammer to avoid "prematures" by the reaction of the vertical hammer, and the guard wire was an additional precaution. The construction of this fuze was kept very secret, only Captain Moorsom and his workman completing them by the insertion of the lead pillar. The first description of these fuzes is said to have appeared in an "Occasional Paper," Royal Artillery Institution, translated by Captain Orr, R.A., from a French account of the fuze, this account having in its turn been compiled from a Russian source.

It will be noticed that in one direction, viz., if the shell strikes bottom foremost, there will be no action. This fuze remained in use in the Navy until 1865.

Armstrong Pillar Fuze.—The Armstrong pillar percussion fuze was introduced in 1861 for rifled breech-loading naval ordnance, and was withdrawn 1869.

It consisted of a body of soft gun-metal, Moorsom gauge, closed at the top by a recessed cover. Inside was a pillar of soft gun-metal, pierced like a tube, and with a detonating patch on the top. This pillar was held in position by a white metal guard and regulator round the neck.

On discharge, the guard sheared its three feathers and jammed itself into an enlarged part in the fuze bottom, the pillar being then only held by the regulator, which it crushed on impact with any substance harder than water, exploding the detonating patch and so firing the shell.

Boxer Percussion Fuze.—The Boxer percussion fuze, introduced 1862, had a thin disc of metal in the head with a projecting pin. The disc was flush with the head of the fuze. Below the pin was fixed, in the body of the fuze, a patch of fulminating composition. On striking an object the disc was bent in and the pin pressed on to the fulminating cap, exploding it and thus firing the shell.

Pettman's Percussion Fuze.—The next important fuze was Pettman's percussion fuze, adopted for naval use, 1862, and which will be here described. After one or two patterns the Pettman General Service (G. S) fuzes were adopted for use in 1866, with small-bore, breech-loading, or muzzle-loading guns (fig. 94).

The body and top plug are made of gun-metal. The

cone plug, detonating ball and steady plug are also of gun-metal but of a harder alloy.

The body is conical, tapped throughout to screw into the G. S. gauge fuze hole, and is slightly hollowed out in the centre to allow play for the detonating ball, and also at the base to allow the lead cup to dovetail into the recess when crushed up.

Two slots are cut across the top of the body to enable the fuze to be screwed into the shell. There is a hole in the centre of the base to allow the cone plug to set back. The top plug has a small cup-shaped recess in its centre to

FIG. 94.—PETTMAN'S G. S. PERCUSSION FUZE.

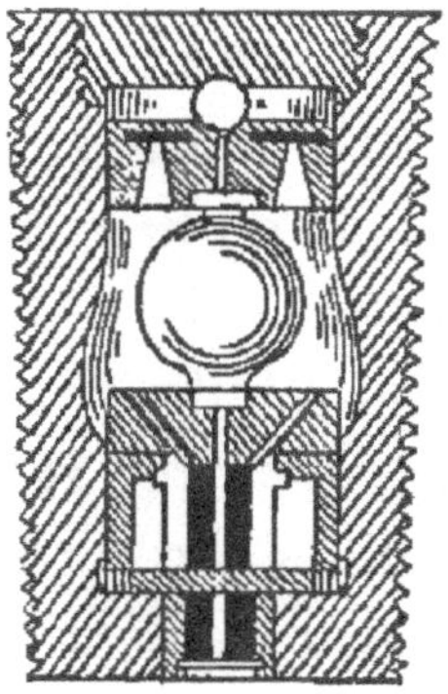

SECTION.

hold the plain ball of brass. The steady plug is a disc recessed at the top and roughened to receive a ring of detonating composition, and having a hollow in the centre to hold the plain ball; three fire holes are pierced to allow the flash to pass down. The bottom of the central hole is enlarged to hold the projection of the detonating ball. A detonating composition is pressed into the recess and covered by a thin copper washer. The detonating ball is roughened by a number of vertical grooves and a deep horizontal groove, and these retain the detonating composition with which the ball is coated. Two projections fit, one in the steady plug and the other in the cone plug. Over the composition two thin copper hemispheres are placed, and the whole is protected by shellaced paper and varnished gut and silk. The cone plug is pierced by three fire holes, the central one holding the base of the detonating ball, and the bottom contains a chamber filled with mealed powder and is recessed near the top of the cylindrical part to allow the lead cup to dovetail on to it. It is pierced near the base for a suspending wire. The lead cup is a hollow cylinder with a flange at the head to fit the recess in the cone plug. There are, therefore, two actions in this fuze, which is designed to act on impact and not on graze.

(1) Suppose a shell fired out of a muzzle-loading gun, the steady plug, ball, and cone plug set back on shock of discharge, the suspending wire is sheared; the lead cup dovetails into the body of the shell and the cone plug, thus preventing its rebound, while the stem of the cone plug protrudes through the base of the fuze, the detonating ball being released from its pivots by the slight wobble of the shell. On striking, the detonating ball is dashed violently against the side of the shell, exploding the composition, and the flash passes through the holes of the cone plug and fires the chamber of mealed powder, and so passes into and explodes the shell.

(2) When fired from a breech-loading gun the steady plug may not disengage, owing to the steadiness of flight of these shells, and in this case the detonating ball will not act. The plain ball is released by the steady plug setting back, and spins round by centrifugal force over the ring of detonating composition on the upper part of the steady plug. On impact the ring is dashed against the plain ball and detonates, the flash passing down through the fire holes and the chamber of the cone plug to the shell.

B. L. Plain Percussion Fuze.—The next fuze, which was a modification of the Armstrong metal concussion fuze introduced in 1860, is the B. L. plain percussion fuze. It consisted of a small plain brass cylinder about an inch long, and was dropped into the shell, after uncapping, and the fuze hole plug replaced or an E. time fuze screwed in. As the principle is similar to the R. L. percussion fuze which superseded it, a description of the latter will suffice for both. These fuzes are designed to act on graze and are still in use. (The former for rifled breech-loading guns only.)

R. L. Percussion Fuze.—The R. L. percussion fuze

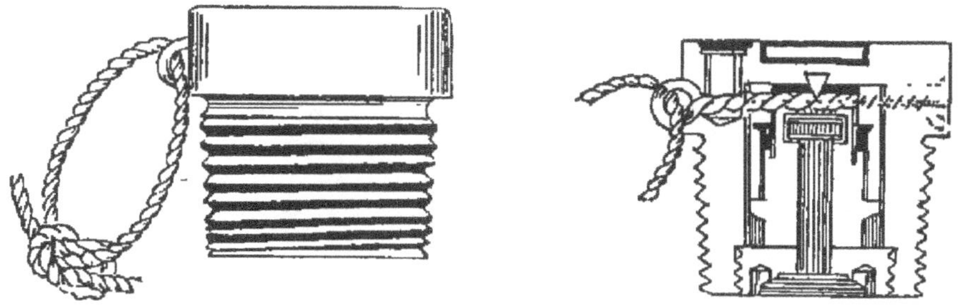

FIG. 95.—R. L. PERCUSSION FUZE. SECTION.

(fig. 95) consists of a gun-metal body, screwed to the G. S. fuze hole gauge, and a bottom plug with a central hole. A steel needle point is fixed in the centre of the head inside the body.

A lead pellet fits inside the body, in the head of which is a cap of detonating composition. On the side of the lead pellet are two small lugs which support the guard and also a recessed V-shaped ring is cut round the pellet above the lugs. The guard consists of a brass collar with a V-shaped reccss, into which the projecting rim of the pellet dovetails. A twisted brass safety pin runs across the top of the body to prevent the accidental contact of the needle and detonating composition. A small vertical hole is also drilled and a lead plug fitted to it from the top of the fuze to seal up the hole left by the withdrawal of the safety pin.

On firing, the shock of discharge sets back the guard which shears off the lead lugs on the pellet, and dovetails itself to the pellet by the lead expanding into the V-shaped groove. The lead plug also sets back and seals the hole from which the safety pin was withdrawn, thus preventing any chance of premature explosion from any flame of the cartridge passing over the shell.

On impact, the pellet and guard fly forward in the body,

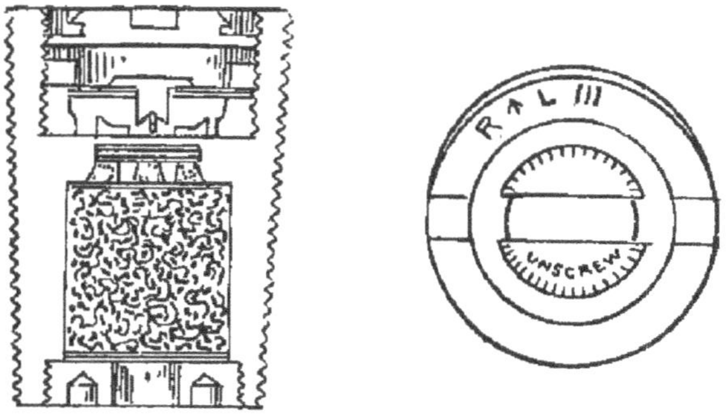

FIG. 96.—FUZE, DIRECT ACTION. MARK III.

and the detonating composition is dashed against the needle, thus exploding, and the flash passes directly through the centre of the pellet and bottom plug into the shell.

Direct-action Percussion Fuze.—The direct-action percussion fuze (fig. 96) was introduced to replace the sensitive fuze in siege train and mountain equipment, but its use has been extended to almost every rifled muzzle-loading and breech-loading gun. It is essentially an "impact" fuze, but will act also on graze if the angle of descent is over ten degrees.

The fuze consists of a body tapped to G. S. pitch (all the way up in the last patterns), which holds in its lower part a blowing charge of fine powder, and the base is closed with a plug with a central fire hole. The upper part is screwed to receive the screw-plug and collar for the needle disc, and above this is screwed a left-handed screw-plug for safety in carrying S.S. shell fuzed (sea service).

The centre part is recessed to receive the detonating composition, below which are nine conical fire holes communicating with the blowing chamber.

The screw-plug for needle disc fits in the upper part of the body, and there is a central hole bored, through which the needle can pass. The needle is supported over this hole by the needle disc of sheet copper, and the disc is kept in position by a collar screwed down above it.

The fuze is prepared by taking out the left-handed safety plug and the fuze remains quiescent until impact with a solid body takes place, when the needle is crushed down on the detonating composition, firing it, and the flash passes through the conical holes to the blowing charge and thence to the shell.

Direct-action Delay Fuze.—A modification of the above fuze called the direct-action delay fuze was similar in construction, but had a longer body, and the flame had to burn through a short length of fuze composition before entering the shell. This delayed the explosion about half a minute, and gave time for the shell to penetrate earth-works, etc. Its use is restricted, however, to land service howitzers.

Small Percussion Fuze.—The small percussion fuze was designed to replace the R. L. percussion fuze, and to act with certainty on graze, and is here described. A number of Armstrong bolt percussion fuzes were ordered for the Navy pending the manufacture of a sufficient quantity of small percussion fuzes; but as no further issue will be made, a description is not necessary.

The small percussion fuze consists of a body screwed

outside to fit the G. S. gauge, bored out from the bottom, and closed with a bottom plug with a central fire hole (fig. 97). A steel needle is screwed through the top of the body.

The detonator pellet contains the R. L. percussion cap, and a chamber filled with three grains of fine-grain powder, and has a slot cut out from one side into which the safety pellet and ball drop. The pellet is prevented from turning in the body by a screw in the latter, projecting into a groove down the side of the detonator pellet. A brass ball prevents the detonator pellet from striking the hammer,

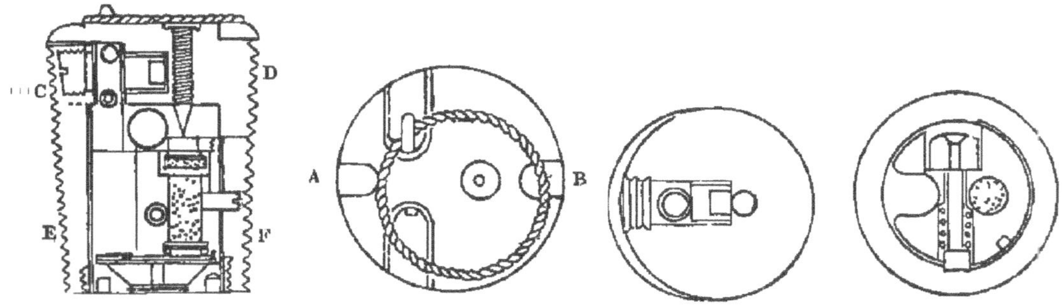

FIG. 97.—FUZE PERCUSSION, SMALL. MARK I*.

and the safety pellet keeps this ball from falling into the slot. The safety pellet is supported by a copper shearing wire and also by the safety pin, and when the latter is withdrawn the hole is closed by a closing pellet of lead and tin. An additional precaution is added in the retaining bolt which passes transversely through the detonator pellet, the head projecting into a recess in the body of the fuze, and kept there by a weak spiral spring.

On loading the shell the safety pin is withdrawn on the shock of discharge, the suspending wire is sheared, and the safety pellet set back to the bottom of the slot in the detonator pellet, the brass ball rolling out and following it.

During flight, the centrifugal force of the heavier end of the retaining bolt overpowers the spring and withdraws the smaller end from the recess, so that the detonator pellet is free to move forward. The lead pellet is driven outwards by the same force and closes the hole left by the safety pin, preventing water or grit from entering the fuze.

On impact the detonator pellet flies forward on to the needle which fires the cap, and the flash ignites the fine-grained powder in the pellet and on the bottom plug, and so passes into the shell.

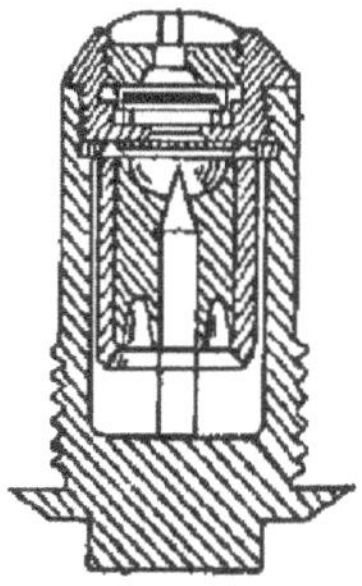

FIG. 98.—HOTCHKISS BASE PERCUSSION FUZE.

With the advent of steel armour-piercing shell, and quick-firing ammunition, a base fuze became a necessity, and there are three types in use; the Hotchkiss, Nordenfelt, and another, designed by the Armstrong Company.

Hotchkiss Base Fuze.—The Hotchkiss fuze is small for a fuze, and consists of a body, percussion pellet, screwed cap screw-plug, and detonating cap (fig. 98).

The percussion pellet consists of a gun-metal casing filled with lead, in which a roughened needle of brass wire is embedded. The percussion cap is contained in the screw cap, and is closed by the screw-plug, having a central fire hole. Before the fuze is in action the detonating pellet is supported by the brass needle. On the shock of discharge it sets back on the needle, the point of which now protrudes through the pellet, and, on impact, pellet and needle fly forward and fire the cap, and the flame passes into the base of the fuze.

Hotchkiss Concussion Fuze.—There is also a Hotchkiss concussion fuze fitted to the head of quick-firing shell, which consists of a wood plug carrying a needle tightly fitting the metal body of the fuze, and the detonating cap is placed at the bottom of this body: there are no safety pins. On impact the head of the fuze is crushed in, and the needle plug forced on to the detonating cap, thus firing it, and the flame passes directly into the shell.

Nordenfelt Base Fuze.—The Nordenfelt quick-firing base percussion fuze (fig. 99) differs from the preceding one in that the needle is fixed and the pellet and detonating composition move. A steel body is fitted with a needle plug of brass. Inside the body a steel pellet, with a cap of

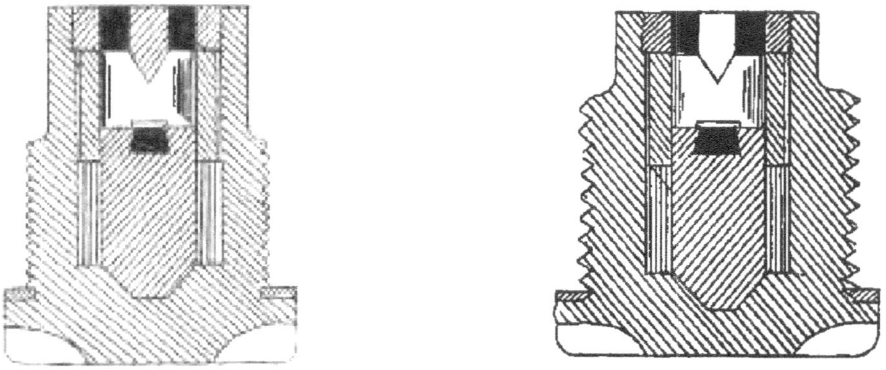

FIG. 99.—NORDENFELT PERCUSSION BASE FUZE.

detonating composition, is kept in position by a coned bearing at the base, and a split ring at the head, which ring grips the upper part of the steel pellet. On discharge the split ring sets down over the steel pellet, and on impact the whole fly forward on to the needle, thus firing the cap and exploding the shell. The Hotchkiss fuze previously described has, however, superseded the Nordenfelt type.

Armstrong Base Fuze.—The Armstrong base percussion fuze (fig. 100) has been introduced for breech-loading and quick-firing armour-piercing shell where it is necessary to keep the pointed head intact. It consists of a gun-metal body screwed into the base of the shell and covered with a lead cap, which is hammered over it to completely seal the fuze hole. A metal pellet fits in the body and contains the detonating cap and a column of pierced mealed powder, to strengthen the flash, in a centre channel. This pellet is kept in position by two retaining bolts, two longitudinal bolts, whose heads project over the top rim of the pellet, and whose extremities are screwed into the base plug, and a spiral brass ring. An annular ring of lead is fitted in the base of the pellet to prevent any jarring or premature

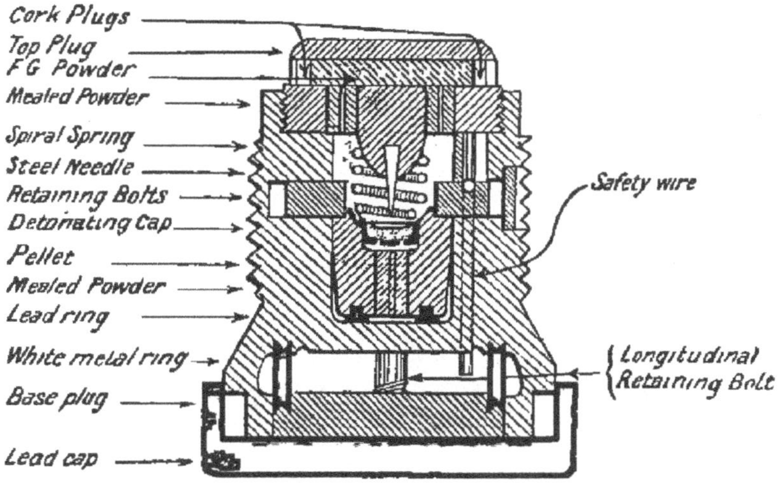

FIG. 100.—ARMSTRONG BASE FUZE.

setting forward of the pellet on to the needle on shock of discharge. The top plug has a needle point fixed in a projecting boss in the centre, and two holes filled with pierced mealed powder at the side allow the flash to pass into a cross channel filled with fine-grain powder, and having its exits into the shell closed with cork plugs. The two retaining bolts are kept in position by wires with their heads bent at right angles, so as to fit into transverse slots cut in the bolts, and with long stems passing down through the body and projecting into a small chamber at the base of the fuze. This lower chamber has a soft metal ring which keeps the base plug of the fuze from being crushed in by accident.

On firing, the force of the powder charge crushes in the base plug, and pushes up the longitudinal bolts and the wires holding the retaining bolts, which latter are now free to fly outwards by centrifugal force, thus releasing the detonating pellet, which is now only kept away from the needle by the spiral spring. On impact the pellet flies forward, overcoming the resistance of the spring, and the needle fires the cap; the flame is strengthened by the powder in the centre channel of the pellet, and flies forward through the holes in the top plug, igniting the powder in the top chamber, which blows off the cap, and so explodes the shell.

We now come to the third series, the time and percussion combined fuzes, which are of more recent dates, and are mainly composed of combinations of time and percussion fuzes previously described.

Armstrong Time and Concussion Fuze.—The Armstrong time and concussion fuze was the pioneer of this class (fig. 101), and was made in three sizes, of which the medium will be here described, and was used with all breech-loading guns.

This fuze is always carried in two separate parts, the body, and the thimble by which the time composition was ignited. The fuze proper consists of a body, collar and nut, two needles, a concussion pellet, guard, split ring, and base plug. The thimble contains the detonator pellet and suspending wire. The body of the fuze is made of a compound of lead, tin, and antimony, threaded on the exterior to G. S gauge, and the bottom is bored out to take the percussion arrangement. Above the screw the diameter is enlarged, and carries the time composition ring. A central stalk holds the time needle, retained in its place by a cross piece, leaving plenty of room for the flash to pass down to a radial channel connecting with the composition ring. The scale is marked in inches and tenths outside. A small hole also communicates, *viâ* a small magazine of pierced mealed powder, with the concussion arrangement. The collar covers the composition ring, and has at one part a small bag of mealed powder, communicating with a ring of mealed powder round the inside of the collar, which serves to convey the flash from the composition ring to the

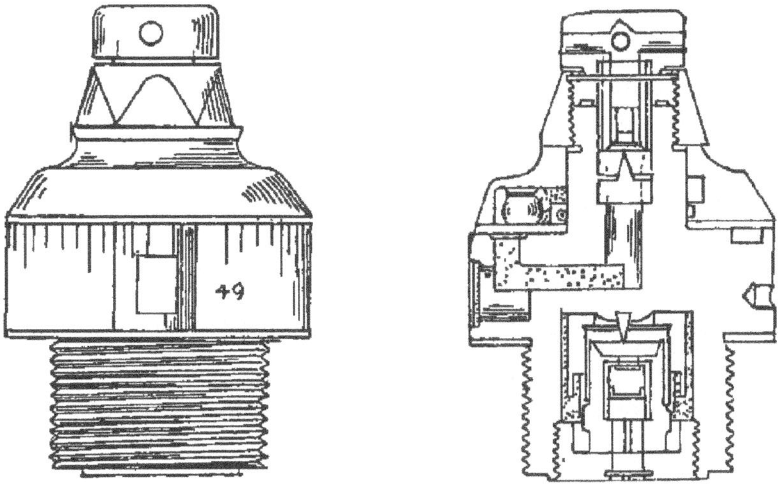

FIG. 101.—(MEDIUM) ARMSTRONG TIME AND CONCUSSION FUZE.

SECTION.

magazine and shell. The collar can be turned round when loosened, so that an arrow head can be brought opposite any required graduation on the body. A clamping-nut secures the collar to the body, and into this nut screws the thimble.

The thimble carries the detonating pellet suspended by a thin brass wire.

The concussion, or graze portion of the fuze, consists of a needle fixed in the body, a concussion pellet containing a percussion cap and powder, a split ring, and a guard resting on the split ring, which latter rests on an enlarged diameter of the pellet.

The thimble is screwed in after the fuze is set, and at the moment of loading.

The action is as follows :—(Time) on shock of discharge, the suspending wire in the thimble is sheared, and the pellet, dashing against the needle, fires the composition ring, which burns round till it comes opposite the small bag of powder, which it ignites, and the flame passes on through the quick-burning ring and down the vertical channel into the concussion portion, and so into the shell. Or (concussion) on discharge the split ring expands, and sets back over the increased diameter of the pellet. At the same time the thin edge of the pellet dovetails into an undercut groove in the guard, and all three are thus locked together. If the concussion part is not fired by the time portion, on impact the guard and pellet fly forward on the lower needle which fires the cap, and the flame passes into the shell through the bottom plug, as in the time portion. Holes are bored in the thimble to allow the gases of combustion to escape.

Time and Percussion Middle Fuze.—The next fuze, which is now in general use, is the time and percussion fuze, and two natures of this fuze are in present use, the "short," burning about nine seconds, used with the smaller guns, and the "middle," burning about fifteen seconds, used with the guns of 6-inch calibre and upwards. As these fuzes are similar in construction, it will be sufficient to describe one of them.

Time and Percussion Short Fuze.—The time and percussion short fuze consists of a body, composition ring, dome, and nut, somewhat similar to the sensitive time fuze, all of which are of gun-metal. There are two safety pins of copper wire, one for the time, and the other for the percussion part of the fuze.

The bottom part of the body is screwed to fit the G. S. gauge, and contains the percussion portion of the fuze, which is so nearly identical with that of the "small percussion fuze" described previously, that it is unnecessary to repeat it.

Above this the body is of larger diameter, fitting over the nose of the shell, and a leather washer makes a tight joint. Above this again the body terminates in a stem, with a central hole forming a gas escape, with holes communicating with the interior of the dome.

The dome fits over the composition ring, and is of metal, stamped up to shape and unturned, and the dome and ring are held to the body by the hexagon-headed nut screwing on the stem. The composition ring contains the detonating arrangement, and the ring of fuze composition in a channel near the periphery. The detonating arrangement consists of a small chamber containing a hammer and needle, supported by a thin copper shearing wire and the safety pin. The ring of composition is blocked at one part so as to burn in one direction only, and in the body of the fuze is bored a hole, communicating between the composition ring and the percussion portion of the fuze. The ring of composition is lined with asbestos, and has a thin leather washer between it and the body, and a hole over that leading to the percussion portion insures the flash passing down.

The ring is graduated outside in eighteen chief divisions to read with an arrow-head marked on the body of the fuze, and corresponding to half seconds of time of flight. These divisions are further subdivided.

The time action is as follows:—The fuze being set by clamping the nut when the ring is turned to any required graduation opposite the arrow-head, and the time safety pin withdrawn, on discharge the hammer shears the suspending wire, and the needle ignites a detonating cap, which fires the composition ring. This ring burns round until it comes to the hole communicating with the percussion portion of the fuze, and the flame passes down, igniting that portion, and exploding the shell.

If required to act as a percussion fuze, only the percussion safety pin is withdrawn, and the fuze acts exactly as in the small percussion fuze; or both pins may be withdrawn, and the shell will act on graze if the time portion does not act previously.

TABLE OF FUZES FOR PRESENT USE.

Fuze.	Nature of Shell	Guns
Pettman's G.S.	Common shell for all guns loaded by hydraulic power .	M.L
Direct Action (D.A.) .	Common, double, and shrapnel 64-pr. and upwards, except when loaded by hydraulic power	M.L.
	Common and shrapnel 4″ and upwards	B.L.
	Common and shrapnel 6″ and shrapnel 4·7″.	Q.F.
Royal Laboratory (R.L.)	Common 7-pr. and 9-pr. . .	M.L.
Middle Time and Percussion or Middle Sensitive Time . . .	Common and shrapnel 4″ to 16·25″	B.L.
	Common and shrapnel 6″ and shrapnel 4 7″	Q.F.
15 Sec. Wood Time without Detonator . . .	Shrapnel shell for all guns (except 9″ when using gas check)	M.L.
15 Sec. Wood Time with Detonator	Shrapnel 9″ when using gas check (on no account to be used in guns loaded by hydraulic power)	M.L.
Armstrong Base Percussion	Armour-piercing steel common and cast iron common 4·7″ .	Q.F.
Hotchkiss Base Percussion	Steel and common 6-pr. and 3-pr.	Q.F.

Note.—Wood time fuzes are never to be used with common shell D.A. Mark III fuzes may be used in muzzle-loading guns loaded by hydraulic power if the plug is left in the fuze.

A few remarks on the methods of firing guns may be here inserted. The linstock, a sort of lighted torch, was in turn followed by the flint-lock, percussion hammer, and detonating tube. As ordnance improved, and rifled guns became the chief armament of ships, the friction tube evolved itself. It is made of copper or quill, but the latter is for naval use, owing to the danger of copper tubes flying about, and cutting the faces and bare feet of the gun detachments.

There have been several patterns of friction tube, but a general description will here suffice for both quill and copper tubes.

The quill friction tube, Mark IV, consists of a body of quill about 2 75 inches long, filled with tube composition, and a roughened nib piece is fitted to pull out of the head, and is smeared with detonating composition. To support the head of the tube, and prevent it from being bent when the nib piece is withdrawn, a leather loop is fastened and woolded to the head with fine copper wire, and this also serves to secure the nib piece in its place. The leather loop is hooked over a small pin screwed into the gun, at such a distance from the vent, that, when the tube is inserted in the vent, the loop will just be long enough for this purpose, and the hook of a lanyard is hooked into the nib piece. A sharp pull on the lanyard withdraws the nib piece, exploding the detonating composition, and firing the composition in the tube, which flashes down the vent and ignites the cartridge in the gun.

With the advent of axial vents and breech-loading (interrupted screw) guns of large calibre, the erosion or "scoring" of the vent by the powder charge was so great that vent-sealing tubes became a necessity, and these are now used for nearly all guns in both services. The principle was similar to the quill friction tube, but tubes are now made for either friction, percussion, or electric firing according to the requirements of each particular lock.

The friction vent-sealing tubes generally consist of a brass case to fit the vent, about $2\frac{1}{2}$ inches long, with a solid head through which a stout wire runs. The inner end of this wire carries the friction roughened piece surrounded with detonating composition, and the outer end is rounded into an eye for hooking the lanyard to, and the tube is filled with pistol powder. The action is similar to the quill friction tube.

The percussion tubes are made for use with the percussion locks, and are similar in form to the friction tubes, but a cap and anvil is fixed inside, and the hammer of the cock falling on the head, or in some cases on a striker, detonates the cap and so fires the tube.

In the electric tubes the wires are led through the head and connected by a fine platinum wire, which forms a bridge inside the tube, and surrounded by loose gun-cotton shreds. On the current passing, the platinum bridge is heated white hot, firing the gun-cotton, which again ignites the pistol powder in the body of the tube, and the flash passes on to the cartridge.

The brass body of these tubes is made thick at the breech end and thinner at the front end, and the explosion of the pistol powder in the tube and cartridge in the gun expands the case to fit tightly the vent, and so prevents any escape of gas.

CHAPTER XV.

THE DEVELOPMENT OF THE MODERN BATTLESHIP.

The idea of plating ships with iron, or at least of carrying the idea out, belongs to Napoleon III., late Emperor of the French. The effects of shells on wooden ships at Sinope and the first bombardment of Sebastopol showed clearly, that unless some means of protection could be devised, ships could no longer hope to attack heavy shore batteries with any chance of success. The result was that both England and France built in 1855 a certain number of what were called floating batteries. These vessels were 172 feet long, 43 feet beam, about 2,500 tons displacement, and drew about 7 feet 9 inches of water; they had engines of 200 horse-power, and could steam about five knots; as ships they were useless, being very unmanageable, but they carried a very heavy armament, viz., sixteen 68-pounders in their batteries, and they were protected by 4 inches of iron on 20 inches of wood, with a bullet-proof iron shelter for the helmsman, from which has sprung the modern conning tower. None of the English floating batteries were completed in time to be of any use in the war with Russia, but three of the French arrived in the Black Sea in time to take part in the attack on Kinburn on October 17th, in which they played a decisive part. Anchored some 800 yards from the main fort, their heavy guns played with crushing effect upon the Russian works, while the storm of shot and shell showered

West & Co., Southsea, photo.

To face p. 294.

BATTALION DRILL. FORMING SQUARE.

upon them by the Russians produced absolutely no effect. Although each of the three vessels was struck some sixty or seventy times, in no case was there more than a dent, from one to one and a half inch deep, and the only casualties occurred through shot and splinters entering through the port-holes. "Everything may be expected from these formidable engines of war," wrote the French Vice-Admiral Bruat in his report.

The immense value of armour having thus been proved, the French Admiralty, and our own somewhat more tardily, began the reconstruction of the respective fleets of the two countries; and in the spring of 1860, the celebrated *Gloire*, the first ironclad battleship, was launched at Toulon. This ship was built of wood and plated entirely with 4·5 inches of iron to 6 feet below the water-line; she was 250 feet long, 55 feet beam, and steamed at a speed of 13·5 knots. The *Gloire* was followed by the launch, January, 1861, from the Thames Shipbuilding Company's works, of the *Warrior* for the English Navy. She was built of iron, but was armour-plated with 4·5 inches of iron for only two-thirds of her length, her bow and stern being unprotected. The *Warrior* and her sister ship the *Black Prince*, completed some months later, were followed by a number of ships in which the armour was carried completely round the hull from some four feet below the water-line up to the level of the upper deck, many of these ships were wooden line-of-battle ships cut down and converted into armoured frigates, while others, like the *Minotaur* and her sisters, were built as armoured ships and had iron hulls. All these earlier battleships, as we call them now, were merely the old frigates, carrying somewhat heavier guns than their wooden sisters on their broadsides, and protected by armour varying from 6 inches at the water-line to 4 inches on other parts of the hull, but otherwise presenting no novel features, either in the design of the ships themselves, or in the method of carrying their armament. But all this time the Civil War in the United States was raging, and on March 9th, 1862, there was fought in Hampton Roads the first naval action between armoured ships, which practically sealed the fate of armoured frigates of the earlier type almost before they had in many cases left the stocks.

When Norfolk, with its dockyard, was evacuated by the Federal troops at the outbreak of the war between the Northern and Southern States, the *Merrimac*, a large 60-gun frigate was set on fire to prevent her falling into the hands of the Confederates; she sank apparently before much damage had been done to her, and when the Confederates occupied the town, a few days later, they succeeded in raising her, when they found her machinery in good repair, and the hull, except where injured by fire in the upper works, sound. Painfully conscious of their weakness at sea, and having in mind the armoured vessels which were being built in Europe, the Confederate authorities determined to attempt the construction of an armoured ship which could cope successfully with the numerous wooden frigates of the Northern fleet.

Under the superintendence of Commander Brooke, and Mr. J. Porter, Constructor of the Confederate Navy, the entire upper works of the *Merrimac* were removed, and she was cut down to the water-line; a rectangular casemate 170 feet long was then built amidships, extending from 2 feet below the water-line to a height over the gun-deck of 7 feet, the sides of which were made of 20 inches of pine, with an outer layer of 4 inches of oak, on which was secured two thicknesses of armour plating, rolled at the Tredegar Works, and manufactured from railway rails, the first tier being put on horizontally and the second up and down; the sides of the casemate were sloped at an angle, according to some accounts, of thirty-five degrees; according to others, the angle was as much as forty-five degrees, and the armour extended some two feet under water, projecting slightly from the hull; both ends of the shield were rounded, so that the pivot-guns could be used as bow and stern-chasers. The roof of the casemate was formed by a grating for ventilation, and also used as a promenade, through which only came her funnel, as the ship had no masts, and at the forward end of the casemate was a small pilot-house protected by 4 inches of iron. She carried ten guns in her battery, one 100-pounder Armstrong gun on a pivot at each end of the casemate, and one 7-inch rifle with three hollow-shot 9-inch Dahlgren guns on each broadside. As her deck ends were two feet below the water and not awash, and she had no masts and rigging, the *Merrimac* was now practically invulnerable to all but the heaviest ordnance. On March 8th, 1862, she steamed out to attack the Federal blockading squadron, consisting of two sailing frigates, the *Cumberland* and *Congress*, and three steam frigates, which latter, however, were unable to come up in time to take part in the action. The *Cumberland* was sunk and the *Congress* had to surrender, the *Merrimac* herself sustaining but little injury, although she sunk the first-named ship by ramming. The next morning she again came out to attack the remaining ships, but was met by an antagonist which, although much smaller, proved quite a match for her; this antagonist was the celebrated *Monitor*. Although it was well known at Washington that the Confederates were turning the *Merrimac* into an armour-clad, it was not until four months later that Mr. Ericsson, the well-known engineer, persuaded the Navy Board to construct a vessel after his own design. The new vessel was to be built at his own risk, and was not to be accepted unless she proved a success; she was further to be completed in the short time of 100 days, so as to be ready to meet the *Merrimac* before the Confederate ship could do any damage; she had to be invulnerable, carry some heavy guns, and yet be of light draught. Ericsson therefore decided to have a turret-ship, as he would thus be able to concentrate the armour over a small space, instead of distributing it upon the larger area of a casemate, while as the turret revolved, the guns in it would command a wide arc of fire. The idea of mounting guns in turrets was not a new one. The late Captain Cowper Coles, R.N., after the Russian war, had submitted designs for such vessels, while the Danish authorities had commenced the construction of such a vessel, but the credit of first carrying out the innovation successfully undoubtedly belongs to Ericsson and the Washington authorities. When complete the *Monitor* was practically a steam raft with a ship's bottom, carrying amidships an armoured turret, in which were mounted two 11-inch Dahlgren smooth-bore guns, firing projectiles of about 150 lb. in weight; she had a displacement of about 1,000 tons, was 172 feet long, with a beam of 41 feet, while she only drew $10\frac{1}{2}$ feet of water. The hull was almost entirely submerged, and when in trim for battle was only 2 feet out of the water, her sides being protected by five 1-inch plates, while the deck was protected by two half-inch plates. The turret, 20 feet in diameter and 9 feet high, protected by eight layers of 1-inch plates, was almost in the centre of the ship, and was turned by steam, revolving on a central pivot, which was supported upon the ship's bottom; the roof was of rolled iron, with ventilating gratings and sliding hatches. On the deck forward was a small armoured pilot-house, which barely held three men with the steering-wheel; abaft the turret were the funnel and two ventilating shafts, the ventilation being entirely artificial. Such was this remarkable vessel, and, as already stated, she proved a match for the *Merrimac;* and although the battle between the two vessels was a drawn one, yet the effect was to save the Northern fleet and effectually dash to the ground the hopes of the Confederates founded on the results of the previous day's fighting. I have gone into some detail of these two ships, partly because the action between them was the first action between armoured vessels, but principally because the result of the action, indecisive though it was, really settled the question of "broadside *versus* turret" in favour of the latter. The step between the *Majestic* of the present day, and the *Monitor* of 1862 is a very long one, in fact it is difficult to realize anything in common between the insignificant-looking *Monitor*, with her single turret barely floating above the surface of the water, and the modern imposing-looking battleship, with her high freeboard bristling with guns; nevertheless, the splendid barbettes on the centre-line of the ship, in which the heavy guns of the modern battleship are mounted, are simply a development of the turret principle, the value of which was so fully proved in Ericsson's experimental vessel on that calm March morning in 1862.

Great was the excitement caused when the news of the action reached Europe, and nowhere more than in England, where a demand immediately arose for the conversion of our fleet into turret-ships. The Admiralty, however, did not move hastily in the matter, but they finally decided to give the plan brought forward by Captain Cowper Coles, for converting the wooden line-of-battle ships into turret-ships, a trial. Accordingly, a screw three-decker, the *Royal Sovereign*, was cut down, armour-plated and fitted with four turrets, each carrying two 9-ton guns, which worked on roller bearings, Captain Coles's own idea, instead of with a central spindle, as fitted by Ericsson to the *Monitor*. She had a freeboard of 6 feet, thus making her a great deal more seaworthy than the *Monitor*, but as the hull of the ship had not been designed to carry the heavy weights now superimposed upon it, she laboured under some disadvantages. Nevertheless she was considered for a time the most formidable ship in the Navy. Attempts have been made to claim for Mr. Ericsson all the credit for the invention of the turret-ship, but in this connection it is only fair to point out, that during the Russian war Captain Coles constructed a domed turret, which was erected on a raft and carried one 68-pounder; moreover, he read a paper on the subject before the Royal United Service Institution in 1860, more than a year before Ericsson commenced the construction of his vessel for the Northern government. A double-turreted monitor called the *Rolf Krake* was also commenced for the Danish government in 1861, which afterwards did good service in 1864 against the Prussian batteries at Eckernstünde, in the war between Denmark and Austria and Prussia. Some foreign countries at this time, notably Russia and the Netherlands, laid down a considerable number of single and double-turreted vessels for coast defence on the model of Ericsson's *Monitor*, and the name was generally adopted as representing this particular class of ship. But in this country some years elapsed before the system of mounting the main armament of ships on the broadside was given up definitely in favour of the turret or barbette. Nevertheless, the advantages of the turret over the broadside system were evident; the fact of the turrets or barbettes being placed in the centre of the ship enables the weight of the guns and the armour to be more evenly and systematically distributed, and it became possible for ships to carry the heaviest guns, which by no possibility could ever have been mounted on the broadside; much greater protection also is afforded by turrets or barbettes both to the guns and their crews, as a greatly increased thickness of armour can be carried on them, and the rounded surface of the turret, or sloping surface of the barbette, will probably cause many projectiles to glance off, which might penetrate if they struck fair.

Although we still continued to construct broadside ships, the increased size of the guns and the corresponding necessity of increasing the thickness of the armour, rendered a change in the design inevitable, and under the auspices of Sir E. Reed, at that time Chief Constructor of the Navy, was commenced the construction of a class of

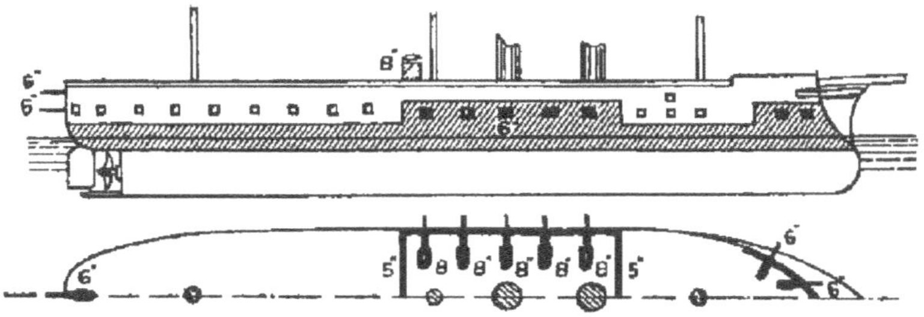

FIG. 102.—PROFILE AND DECK PLANS OF H M.S. "BELLEROPHON," SHOWING DISPOSITION OF ARMOUR AND GUNS.

ship, known as central or box-battery ships, the first of which, the *Bellerophon* (fig. 102), was completed in the early part of 1866. She is a ship of 7,550 tons, 300 feet long, with a beam of 56 feet. Instead of being completely covered with armour, and having her guns distributed along the whole length of the broadside, it will be observed that she has a complete water-line belt, which is carried up over a battery in the centre part of the ship, while before and abaft this battery the ship's hull above the belt is unprotected; the thickness of the armour on belt and battery is 6 inches, and the ends of the battery are shut in by 5-inch armoured athwartships bulkheads. She originally carried twelve 12-ton muzzle-loading guns, but some years ago was re-armed with ten 8-inch breech-loaders. As has been already mentioned, it is to Elswick and Rear-Admiral Scott that the country is indebted for the iron carriages and slides, with their patent compressors for checking the recoil, which made the mounting of such heavy guns on the broadside possible. The construction of central-battery ships was continued for some ten years after the completion of the *Bellerophon*, various improvements being made so as to combine some of the advantages of the turret-ship with the broadside system of mounting guns, by giving an end-on fire to the foremost and after battery guns by means of recessing the sides of the ships, and so increasing the arc of training; the two last ships of this class, the *Alexandra* and *Téméraire* being completed in 1877. In these two last-named magnificent ships the thickness of the armour increased to 12 inches on the water-line belt, and 8 inches over the battery, the athwartships bulkheads being 8·5 inches thick. The *Alexandra*, when she was first commissioned in 1877, mounted in her main or lower battery six 18-ton muzzle-loading rifled guns, three on each broadside, and two 25-ton muzzle-loading guns, which were shut off by an armoured athwartships bulkhead from the other guns, with an arc of fire from right ahead to the beam on each side of the ship respectively; in the upper battery two 25-ton and two 18-ton muzzle-loading rifled guns were mounted, the two 25-ton guns firing from right ahead to the beam, and the two 18-ton guns firing from the beam to right astern (fig. 103). The ship now carries in her upper battery four 9·2-inch 22-ton breech-loading guns, and two 10-inch muzzle-loading guns have been substituted for the two 25-ton guns in the foremost main-deck battery.

The *Téméraire* has a main battery similar to the *Alexandra's;* but she only mounts four 18-ton guns instead of six on the broadside, with two 25-ton guns in a separate armoured casemate, which, as in the *Alexandra*, have an arc of fire from right ahead to abeam. She has no upper-deck battery, but instead has two 25-ton guns, mounted on disappearing carriages, to fire *en barbette*, one forward and one aft; these barbettes are protected by 10-inch armour, with armoured ammunition tubes to the magazines.

Although the construction of broadside battleships was continued during the period I have just mentioned, turret-ships were not neglected. A limited number of these ships, with low freeboards were constructed for coast-

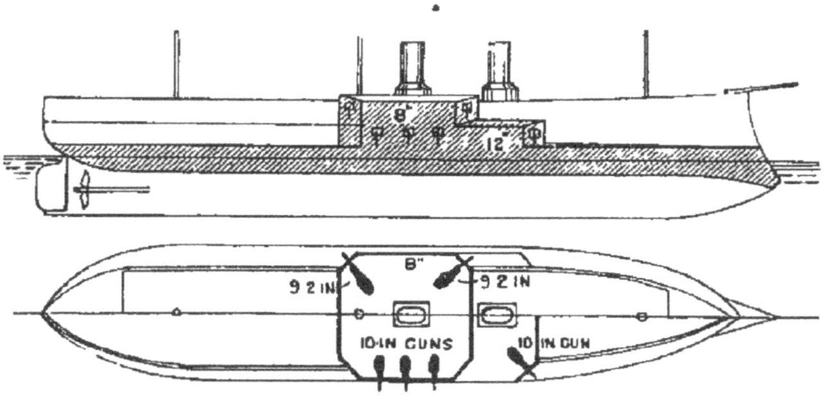

FIG. 103.—PROFILE AND DECK PLANS OF H.M.S. "ALEXANDRA," SHOWING DISPOSITION OF ARMOUR AND GUNS.

defence; two others, the *Monarch*, by the Admiralty, and the unfortunate *Captain*, under the superintendence of Captain Cowper Coles, her designer, were built as sea-going vessels of this type; each had a couple of turrets, in which were mounted two 25-ton guns; these turrets were at each end of an armoured breast-work, which, rising from the water-line belt, extended amidships for about a third of the length of the ship. The difference between the two ships lay in the freeboard, which in the *Captain* was only 6 feet, while in the *Monarch* it was as much as 14 feet. The fate of the unfortunate *Captain* is well known, the *Monarch,* although out of date as a fighting ship, is still on the strength of the Navy. But the three most notable turret-ships were laid down in 1869, viz , the *Devastation, Thunderer,* and *Dreadnought.* They were designed as mastless, low freeboard sea-going turret ships, and although now twenty-six years old, are sufficiently formidable to be classed as second-class battleships. A short description of one, the *Devastation,* must suffice, as for all practical purposes they may be regarded as sister ships, the *Dreadnought* being somewhat larger than the other two. The *Devastation* is a vessel of 9,330 tons displacement, 285 feet long, with a beam of 62 feet; she is protected by a belt of 12-inch armour at the water-line. Above this, running above two-thirds of the length of the ship, is an armoured redoubt or casemate, also with 12-inch armour, which protects the base of the turrets, loading gear, etc.; an armoured deck, resting on top of the belt, covers in the whole ship outside the casemate. The turrets rising at each end out of this casemate are protected by 14-inch armour. The freeboard of these ships is extremely low, the deck being only some four feet above the water, thus necessitating at sea the closing of all apertures by water-tight hatches, as the deck is continually under water. A superstructure or hurricane deck is erected in the space between and slightly above the turrets, where accommodation is found for the boats, chart-houses, and armoured conning tower (fig. 104). At the time of her completion, at the end of 1872, the *Devastation* was unquestionably the most formidable fighting ship in the world, and this in spite of the fact that her extremely low freeboard forward made it difficult, if not impracticable, to maintain a high rate of steaming against a heavy sea; but she was more efficiently and heavily armoured than any other ship afloat at the time. She presented but a small mark to an enemy, while at the same time the two 25-ton guns she carried in each of her turrets, perfectly protected as they were, could all four be brought to bear on any given part of the hull of an enemy from almost any position the ship chose to take up when attacking, except when end on or nearly so. When it is remembered that less than six years before the completion of the *Devastation*, our flagship in the Mediterranean was a three-

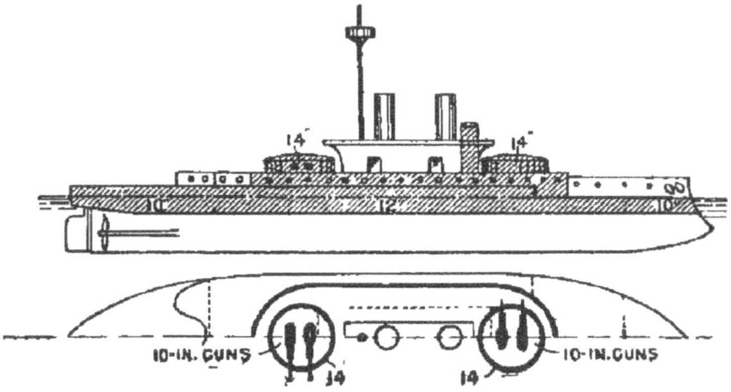

FIG. 104.—PROFILE AND DECK PLANS OF H.M.S. "DEVASTATION," SHOWING DISPOSITION OF ARMOUR AND GUNS.

decker, the *Victoria*, of 121 guns, and that thirteen years before her completion not only was there no single armoured ship afloat (for the floating batteries built at the end of the Russian War can hardly be counted), but whether such ships could be successfully constructed was even a moot point, the complete revolution in naval architecture, as represented by the change from the wooden three-decker of 1860 to the *Devastation* of 1872, is one of the most wonderful and startling, certainly in naval history. The *Devastation* and her sister, the *Thunderer*, have both lately been re-armed, the new 10-inch 29-ton breechloader taking the place of the old muzzle-loaders, so that they are now more efficient fighting ships than when they were first built.

By the time the *Alexandra* and *Téméraire* were completed in 1877, it had become evident that, in view of the new and more powerful breech-loading guns, which it was seen would have to be adopted in the near future, much thicker armour would be required if a ship was to be effectually protected, and as the power of ships to carry more than a certain amount of armour is limited, the important question arose as to how the smaller amount of armour, which its increased thickness would only allow to be carried, should be distributed over a ship, so as to afford the best protection for guns and the vital parts of the vessel, and also as to the best method of monnting the much heavier guns, which would now form the principal armament of large ships. The government therefore appointed a committee of naval officers and experts to consider and report on the best design for the battleship of the future. The result of their deliberation was the adoption of what is known as the "Citadel" type of ship, and on this principle, with certain modifications, several battleships were built. In this type of ship the continuous armour-belt round the water-line is done away with, and in its place the armour, of great thickness, is concentrated round a citadel in the central portion of the ship; the length of the citadel varies from about one-third the length of the ship to nearly a half; the armour extends to a depth of some four feet below the water and about six feet above; at each extremity rise the turrets in which the guns are mounted. In the *Inflexible*, the first ship of this type built, the armour at the water-line is 24 inches thick, with 17 inches on the turrets. The protection afforded to the guns, loading arrangements, and vital parts is in this class of ship excellent, but against this has to be set the great extent of water-line and hull unprotected by any armour at all, as may be seen in the accompanying plan of the *Colossus* (fig. 105), and grave doubts have consequently been thrown on the real value of these vessels as fighting ships, if due weight is given to their manifest defects.

By the year 1880, and before any of the citadel class had been completed, a new enemy to ships had appeared, in the shape of the torpedo-boat, which rendered it necessary for

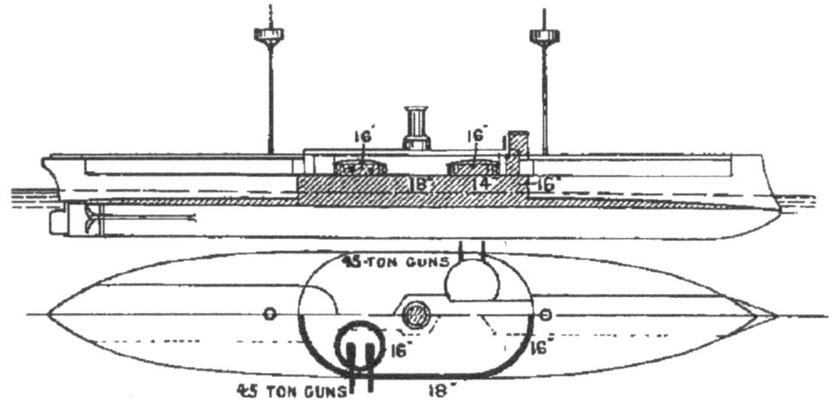

FIG. 105 —PROFILE AND DECK PLANS OF H.M.S. "COLOSSUS," SHOWING DISPOSITION OF ARMOUR AND GUNS.

the heavy armament to be supplemented by a secondary battery of lighter guns; and to the necessity of meeting the attacks of these swift little vessels, we are indebted for the rapid development of the quick-firing system, which has been the marked feature in the improvement of ordnance during the last few years. It thus became necessary once more to carry guns on the broadside, and the design of the citadel ship had to be materially modified. Six battleships were laid down, which are now known as the "Admiral" class, being named after distinguished naval commanders. Much controversy has arisen over these ships; on the one hand they share the defect of the citadel type, in having long unarmoured ends, and they are deficient generally in armour protection; on the other, such armour as they carry is of great thickness, while their armament is a formidable one. The most notable feature in these ships is, that in them the Admiralty for the first time adopted the system of mounting the heavy guns in barbettes. The difference between the turret and barbette systems is this: that where the guns are mounted in turrets, the latter revolve, carrying the guns with them: where the guns are mounted in barbettes, the guns are on turn-tables, which revolve with them inside the barbette, which is itself a fixed heavily-armoured redoubt, the guns firing over the top. The advantage claimed for the barbette system, which up till quite recently has been the French favourite method of mounting guns, is that the guns are carried much higher out of the water; on the other hand, they are much more liable to injury from an enemy's fire, as, except when loading, the whole gun is completely exposed. This defect has, however, been overcome in all our latest battleships of the *Renown* and *Majestic* types, as will be explained later on. The "Admirals" are ships, roughly, of some 10,000 tons, as they are not all exactly alike, the *Collingwood* being somewhat smaller, and the *Benbow* being somewhat larger than the other four. They are 325 feet long, with a beam of 68 feet. A belt of 18 inches of compound armour[1] protects about 150 feet of the water-line amidships; at each end of the belt 16-inch armour is carried up to the level of the upper deck, which protects the loading arrangements of the guns, ammunition, supply, etc.; 16-inch athwartships bulkheads close this citadel in and protect the

[1] Compound armour is iron armour faced with steel.

secondary battery from a raking fire; at each end of the citadel rise the barbettes, in each of which are mounted two 67-ton guns, protected by 11-inch plates, placed at an angle, thus much increasing their resisting power (fig. 106). Between the barbettes is the secondary battery, in which are mounted the six 6-inch guns (ten 6-inch guns in the *Benbow*) which form it, while on the superstructure above are carried twenty-two 6-pounder and 3-pounder quick-firing guns, including those in the fighting tops. When the 6-inch guns in the battery are replaced by

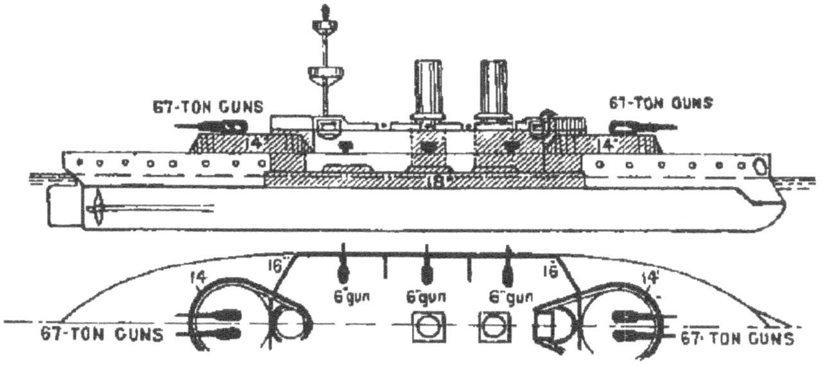

FIG. 106.—PROFILE AND DECK PLANS OF THE "ADMIRAL" CLASS, SHOWING DISPOSITION OF ARMOUR AND GUNS.

6-inch quick-firers, from an offensive point of view, as far as gun-power is concerned, there is little left to be desired in these ships; but the want of proper protection for the 6-inch guns is a grave defect, and the only thing to be said is, that it is a defect equally common to the secondary batteries of the larger proportion of the French battleships.

In view of the strong feeling that the ships of the "Admiral" class were sadly deficient in armour protection, and also of the rapid growth in the size of quick-firing guns, and the development of high explosives for use in shells, a new departure was taken by the Admiralty in 1886 by the laying down of the *Nile* and *Trafalgar*. For their offensive powers, and for the completeness of their armour protection, they show an extraordinary advance over the ships of the "Admiral" class and their predecessors. Although not completely belted all round, they are very nearly so, and they may be described as improved *Devastations*, with the addition of a central battery between the turrets for the auxiliary armament in place of the superstructure of the earlier ships. They are vessels of 11,940 tons, 345 feet long, with a beam of 73 feet; there is a water-line belt of 20-inch compound armour for 230 feet of the whole length of ship, shut in by 16-inch armoured bulkheads; above this belt is an armoured citadel, 133 feet long, and protected by 20 inches of compound armour tapering to 16 inches, with rounded bulkheads of 18-inch armour tapering to 14 inches; this citadel protects the base of the turrets, which rise from it at each end, and are in their turn protected by 18-inch compound armour. Above this citadel is the armoured upper battery, 110 feet long, protected by 5-inch steel armour, in which are mounted six 4·7-inch quick-firing guns, while two 67-ton guns are mounted in each turret (fig. 107); and there are in addition twenty 6-pounder and 3-pounder quick-firing guns.

But the *Nile* and *Trafalgar*, extremely powerful fighting machines as they are, did not entirely embody, in the view of our naval authorities, all the requirements of the perfect first-class battleship. Like all turret ships they have, comparatively speaking, low freeboards, while the turret guns, as has been already pointed out, are much nearer the water than when mounted *en barbette*. This in bad weather is a great disadvantage, as when a ship is steaming against a head sea, the working of the foremost guns would be seriously hampered; so when what is known as the Naval Defence

Act of 1889 was passed, it was decided, with the exception of one ship, the *Hood*, that the new battleships to be constructed under it should be barbette ships. Great pains were taken with their designs, so as to make them a homogeneous squadron of the most formidable battleships as yet constructed in any country. Ten battleships in all were to be constructed, of which two were to be somewhat smaller than the others, to enable them to go

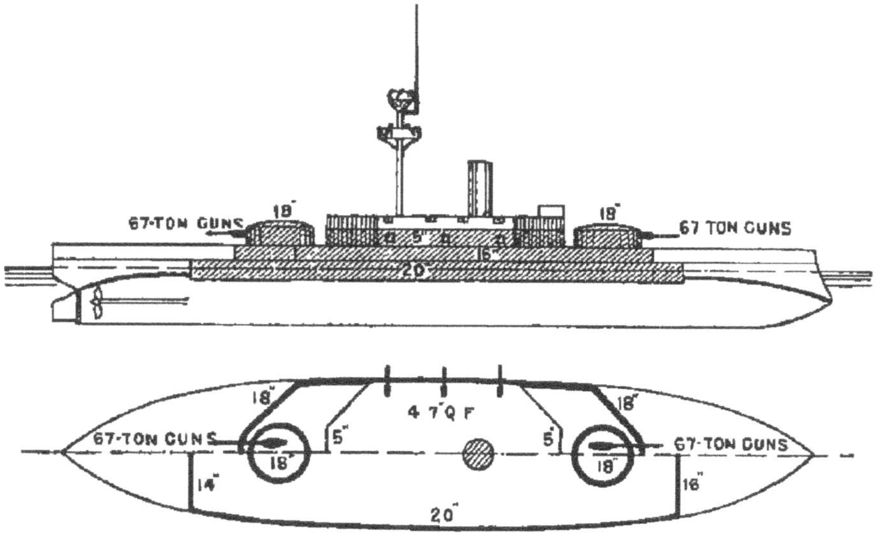

FIG. 107.—PROFILE AND DECK PLANS OF H.M.S. "NILE" AND "TRAFALGAR," SHOWING DISPOSITION OF ARMOUR AND GUNS.

through the Suez Canal. Of the remaining eight, all were of 14,000 tons, far exceeding in weight any ships hitherto built for the Navy, with a length of 380 feet and a beam of 75 feet. Seven, known as the *Royal Sovereign* class (fig. 108), carry their 67-ton guns in barbettes, while the eighth, the *Hood*, carries hers in turrets. All the ships have high freeboards, thus resembling much more in outward appearance the old broadside ships than the original barbette ships of the "Admiral" class. The disposition of armour in all is the same; there is, first, a water-line belt of 18-inch compound armour, $8\frac{1}{2}$ feet deep, running for two-thirds of the whole length of the ship, terminated at each end by 16-inch transverse bulkheads; above the belt is a 3-inch steel deck, and at each end (about 50 feet from the bow and stern respectively) are the armoured citadels for the barbettes, protected by 17-inch compound armour to the height of the spar-deck; the barbettes themselves are similarly protected, while armoured ammunition tubes are carried down from the turn-tables to the top of the belt; this arrangement precludes the possibility of the bursting of shells containing large explosive charges under the floors of the barbettes, upon which the revolving gun platforms are carried. Between the barbettes the broadside above the belt is protected to a height of $9\frac{1}{2}$ feet above water by 5-inch armour, while the four 6-inch quick-firing guns on the main deck are mounted in casemates with armour of a similar thickness; the remaining six 6-inch quick-firers are mounted on the spar-deck, three each side, and are protected by steel shields attached to the mounting, and revolving with the gun. This method of distributing the guns of the secondary battery on two decks was adopted in order to secure the widest possible distribution of the guns in view of the development of high explosives. In the *Hood*, in place of the barbettes of the other ships, there is at each end of the belt a lozenge-shaped redoubt, covered with 17-inch armour, which inclose the bases of the turrets, upper portion of the ammunition hoists, and loading appliances; but it is difficult to understand why the turret system was adopted for this ship, as the barbette guns are 6 feet higher out of the water, and the ships have a superior height of free-board fore and aft, in the proportion of 18 feet to $11\frac{1}{2}$ feet.

Before proceeding to notice the latest type of battleship, viz., the *Majestic* class, which are improved *Royal Sovereigns*, a short description of the *Centurion* and *Barfleur*, the two smaller battleships built under the Naval Defence Act must be given, as they are the first ships in which the heavy barbette guns have been protected by armoured

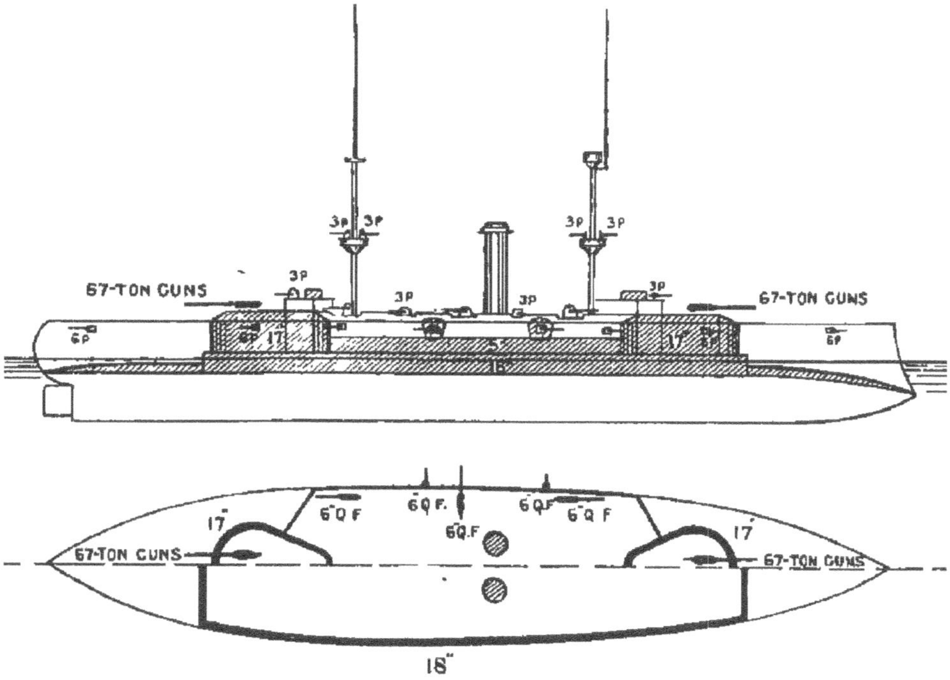

FIG. 108.—PROFILE AND DECK PLANS OF "ROYAL SOVEREIGN" CLASS, SHOWING DISTRIBUTION OF ARMOUR AND GUNS.

hoods. These two ships are 360 feet long, with a beam of 70 feet, and a displacement of 10,500 tons. A belt of compound armour, having a maximum thickness of 12 inches, protects the water-line for 250 feet of their length, and extends below water to a depth of 5 feet, and above, to a height of $2\frac{1}{2}$ feet; above this again is a 4-inch belt of Harveyized steel extending up the side to the main-deck. The heavy armament of four 10-inch 29-ton guns is mounted in redoubts *en barbette* protected by 9-inch armour, which rise from each end of the water-line belt; these redoubts are topped with 6-inch nickel-steel hoods for the protection of the breech of the guns and guns' crews, and rotate with the guns as they are trained on the turn-tables; these hoods are pear-shaped in plan, slope upwards and backwards after the manner of a barbette; an arrangement which gives an effective resistance to horizontal fire of 11 inches, although they possess an actual thickness of 6 inches only, while the obliquity of the armour has also the additional advantage of causing the shot to glance off before penetration. The secondary battery consists of ten 4·7-inch quick-firing guns, four of which are mounted in 4-inch casemates on the main-deck, with 3-inch shields on the guns, and the remaining six being mounted on the upper deck with 4·5-inch shields.

A further development in the protection of guns and guns' crews was taken about this time by four of the 6-inch quick-firing guns, mounted in the first-class cruisers *Blake*, *Blenheim*, and the nine vessels of the *Edgar* class being placed in 6-inch armoured casemates on the main-deck; while in the newest first-class cruisers, viz., the *Powerful*, *Terrible*, and the *Diadem* class, this protection has been carried still further, and the whole of the 6-inch quick-firing guns on the broadside are in 6-inch casemates made of nickel steel, with 3-inch shields on the guns.

We now come to the latest development of the battleship in our Navy, represented by that splendid group of ships, the *Majestic* and her eight sisters, the *Magnificent*, *Mars*, *Prince George*, *Victorious*, *Jupiter*, *Cæsar*, *Illustrious*, and *Hannibal*. These ships are of slightly larger tonnage than the *Royal Sovereign* class, being 390 feet long, with a beam of 75 feet, and a displacement of 14,900 tons; and they show a very large area of side protection, in fact, they may be really described as side-armoured, in contradistinction to the term belted. Instead of the 18-inch belt of compound armour, with another narrow belt of 5-inch armour above, as in the *Royal Sovereigns* the *Majestics* are protected by 9-inch armour of Harvey steel, which extends from 6 feet below the water-line to 10 feet above for two-thirds of their length, while a most important modification

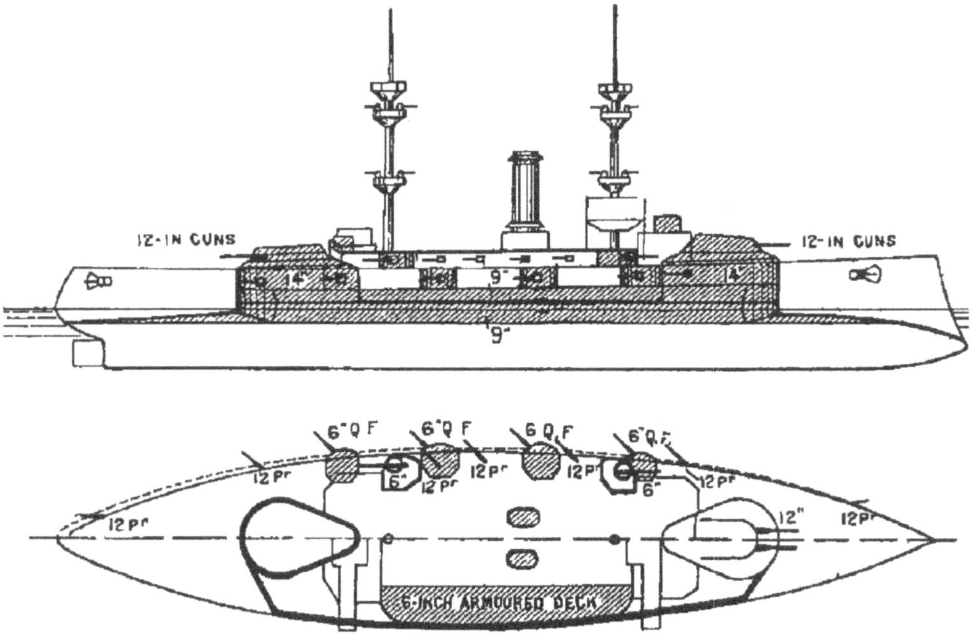

FIG. 109.—PROFILE AND DECK PLANS OF THE NINE SHIPS OF THE "MAJESTIC" CLASS, SHOWING DISPOSITION OF ARMOUR AND GUNS.

has been made in the armoured deck, which, instead of being flat on top of the belt, curves downward on each broadside to the armoured shelf or lower edge of the side armour, and as it is 4 inches thick, in addition to having the side armour to penetrate, a projectile would meet with 4 inches of steel set at an angle, which would really mean 6 inches of steel to pass through. The height of the curve of the armoured deck is about 9 feet, and it extends throughout from apex to apex of the citadel (fig. 109). The two barbettes rise at either end of the citadel and are pear-shaped, with a thickness of armour of 14·6 inches, while the armour bulkheads, inclosing the ends of the citadel are a prolongation of the side armour, increased to a thickness of 14·9 inches, which is carried round thus to meet the barbettes at the central line, and they inclose the curved slope of the armoured deck, being built against it. In each barbette are mounted two of the new 12-inch 46-ton wire guns, which are carried some 26 feet above the water-line, being nearly 4 feet higher out of the water than the *Royal Sovereign's*, and a great advance on the 14 feet of the *Trafalgar's* guns. The heavy guns in all the new battleships are further protected by 10-inch nickel steel hoods, which are made with sloping sides and roof, so that they offer no surface to a direct hit from a projectile. Another improvement in the *Majestic* class consists in all the twelve 6-inch quick-firing guns forming the secondary battery, being in separate 6-inch nickel-steel casemates, with 3-inch shields on the guns.

With the extension of the quick-firing armament to ships, it has been found necessary to increase the thickness of the shields which protect most of the guns mounted in cruisers and on the upper decks of some of our battleships. Some years ago shields 1 inch and 2 inches thick were considered sufficient protection, but now, even for 4·7-inch guns, the shields are made $4\frac{1}{2}$ inches thick in front; while for the 6-inch and 9·2-inch guns the shields are arranged to show in front a thickness of 6 inches. Originally the shields were simply bolted on, but it was proved that under such conditions a blow on the shield was communicated to the mounting, and did almost as much damage to the latter as if there were no shield. The Elswick firm has devised a new arrangement for attaching the shields, which has overcome this objection; the matter was one of great importance, for with a rigid attachment even a small blow from a shot might distort the shield enough to jam and disable the mounting. The shield has to be attached so that there may be some space between it and the mounting; at the same time the attachment should have sufficient elasticity to yield to a blow, and not transmit it entirely to the mounting. Elswick has managed this by hanging in the 6-inch and pedestal mountings the outer shields at about 6 inches distance from the mounting by steel brackets, or stays, which are bent so as to allow of as much elasticity as possible. The inner shield, as has been mentioned in a previous chapter, is cylindrical in shape and 3 inches thick; the outer shield for the 6-inch mountings is now made of 3-inch steel in front, tapering to $1\frac{1}{4}$ inch in rear, with a sloping roof of three-quarters of an inch. The shield is not circular, but of a parabolic shape, and it is hoped that a shell from a quick-firing gun—the projectile against which the shields are specially designed as defence—would be burst on the outer shield, and that if any fragments perforated they would be stopped on the inner shield.

Before concluding this chapter it may be of interest to point out the methods of mounting and protecting the heavy guns adopted in one or two of the principal navies of Europe. In affording protection to their battleships the French have, from the earliest days of armour, persistently adopted the principle of having a complete all-round water-line armour belt, whatever sacrifices this may have entailed in other directions. They have always been extremely fond of mounting their heavy guns *en barbette;* but curiously enough in their latest type of first-class battleships (the *Charlemagne* and sisters), of which there are at present four building, they have given up the barbette system and the heavy guns will be in closed turrets, one forward and one aft, as in our *Hood*, this change being made at the time when we seem to have given up the turret definitely in favour of the barbette.

In the earlier types of French armour-clads are found much the same characteristics as in our own; first, the armoured broadside frigate; then, with increased thickness of armour and heavier guns, came the central-battery ships, but instead of having the upper-deck guns in an upper battery, the two or four heavy guns, as the case may be, were mounted in small barbettes, which were generally sponsoned out, one on each beam, where only two guns were carried on deck; or, where there were four, the other two were mounted in barbettes placed forward and aft on the centre line of the ship. As in the case of our own earlier types, there is considerable diversity to be found among the French ships, but with the exception of their coast defence vessels, all French sea-going battleships, except one, the *Hoche*, are built with high freeboards, and, as has been already mentioned, all, whether sea-going or coast defence, have a complete water-line armour belt. The French naval authorities recognized before we did, moreover, the advisability of supplementing the heavy armament with a considerable number of lighter guns; thus, at the time we were building our *Ajax*, *Agamemnon*, *Inflexible*, *Colossus*, and *Edinburgh*, practically as turret-ships pure and simple, we find contemporary French ships with secondary batteries of from eight to fourteen medium-sized guns; our "Admiral" class being the first vessels in which we attempted to have a secondary armament in addition to the heavy barbette guns. The heaviest gun in the French navy is the 42 cm. (16·5 inch) gun, two of which have been mounted in hooded barbettes, one forward and one aft, in the four coast defence battleships, *Caiman*, *Indomptable*, *Requin*, and *Terrible*. These guns weigh 75 tons, and throw a 1,719 lb. projectile, with a muzzle velocity of 1,739 feet, capable of perforating 27·4 inches of iron. Another type of this gun with a calibre of 37 cm. (14·5 inches), throwing a 1,235 lb. projectile, with a muzzle velocity of 1,969 feet, is mounted in the two first-class battleships, *Amiral-Baudin* and *Formidable*, each of which carries three singly in barbettes, protected by 16-inch armour, one forward, one amidships, and one aft (fig. 110), besides a secondary battery of twelve 14 cm. (5·5-inch) quick-firing

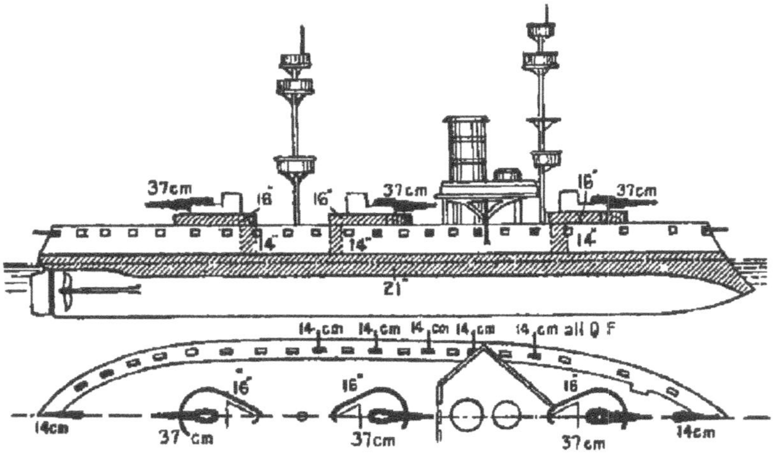

FIG 110.—PROFILE AND DECK PLANS OF "AMIRAL-BAUDIN" AND "FORMIDABLE," SHOWING DISPOSITION OF ARMOUR AND GUNS.

guns, with thirty-two small quick-firing and machine guns, and six torpedo tubes. The armoured water-line belt is 21 inches thick, tapering away at bow and stern to 10 inches, but the weak point about these ships, which is to be found also in later types, is the want of proper protection to the base of the barbettes; there is only a small armoured ammunition tube, and it is quite conceivable that a shell bursting under the turn-table would suffice to effectually disable one of these guns; this, as has been pointed out already, is one of the defects in our own "Admiral" class, a defect, however, which has been effectually remedied in all our battleships built since. The secondary batteries in these ships are quite unprotected, while from their high freeboard they offer an enormous target.

The next battleship constructed after the two just described is the *Hoche*, noticcable as being the only sea-going French battleship with low freeboard fore and aft; in armour protection, arrangement and weight of armament, she is, for practical purposes, identical with the three which followed her, the *Marceau*, *Magenta*, and *Neptune*. All these vessels have an 18-inch complete water-line steel belt, tapering to 10 inches at bow and stern, with 16-inch armour on the barbettes, and 6-inch armoured ammunition tubes to the gun positions. The armament consists of four 13·3-inch guns, which throw a projectile 881 lb. in weight, with a muzzle velocity of 2,625 feet; they are mounted singly in barbettes, one forward, one aft, and one on each beam, the barbettes on the beam being sponsoned out to allow of these two guns firing fore and aft, either ahead or astern. The secondary battery consists of sixteen 14 cm. (5·5-inch) quick-firing guns, with thirty-four small quick-firing and machine guns (fig. 111). The above method of mounting heavy guns has never found favour with English gunnery experts, or our naval authorities at Whitehall, although several well-known critics outside the Navy strongly urged at one time the advantages of the French system, on the ground of the importance of having as strong a bow and stern fire as can be secured. Now although a powerful bow or stern fire is of importance to cruisers, as they will most probably be in the position of either chasing or being chased, in battleships, on the other hand, it is the weight of the broadside fire which will, as in the days of old, decide most actions, and not the end-on fire. The French ships, with their guns arranged as described, can bring, it is true, three of their heavy guns to bear on an object when they are directly end-

West & Co., Southsea, photo

To face p. 320.

FIELD-GUN DRILL.

on to it, but a swerve of a quarter of a point will suffice to throw one of the heavy guns in the beam barbettes out of fire completely; while, when it comes to fighting broadside to broadside, they can only bring three of their heavy guns to bear on an enemy's ship. In our ships, on the other hand, with the heavy guns mounted in pairs, in two barbettes, although only two can be brought to bear at a

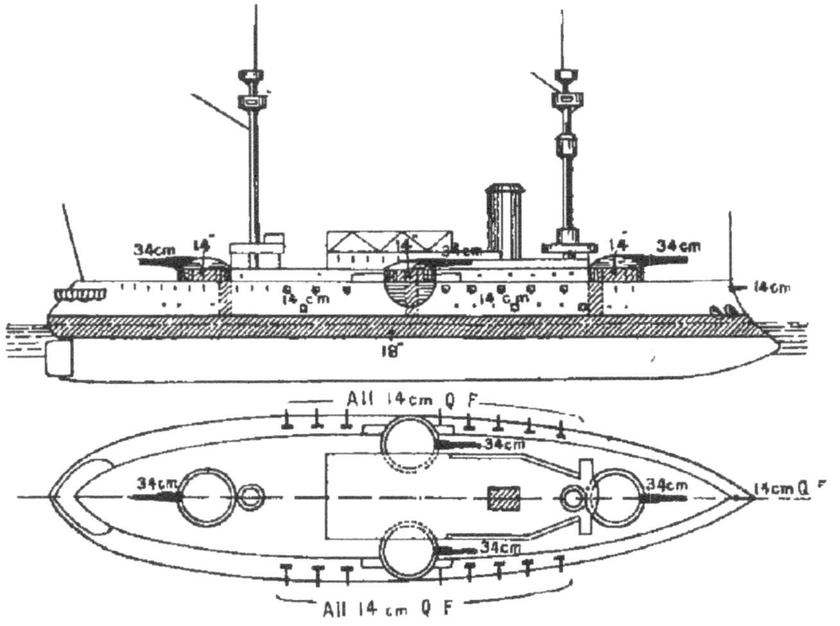

FIG. 111.—PROFILE AND DECK PLANS OF "NEPTUNE," "MARCEAU" AND "MAGENTA," SHOWING DISPOSITION OF ARMOUR AND GUNS.

time on an object ahead or astern, yet those two guns can be kept steadily bearing on the object, no matter how much the ship's head may fall off one way or the other; while, instead of only three, our ships can bring all four heavy guns to bear on the broadside, a matter of supreme importance, as the concentrated fire of the four guns, against three firing from different positions, might well determine the fate of an action. That the French have at last begun to recognize this is evident, from the fact that in all their new ships the heavy guns are, for the first time, being mounted in pairs in turrets.

After the *Marceau* and her sisters we come to a group of five ships, which are the most powerful yet built for the French Navy; these are the *Bouvet*, *Carnot*, *Charles Martel*, *Jauréguiberry*, and *Massena*. None of these ships are as yet ready for sea, although two, the *Jauréguiberry* and *Charles Martel*, are undergoing their trials. Their armament consists of two 12-inch guns, mounted in barbettes, one forward and one aft, which throw a projectile of 626 lb. in weight, with a muzzle velocity of 2,625 feet, and of two 10 8-inch guns, one on each beam, in a sponsoned-out barbette; these guns throw a projectile of 476 lb. in weight with a muzzle velocity of 1,969 feet; the secondary armament consists of eight 14 cm. (5·5-inch) quick-firing guns, mounted in pairs in small turrets (fig. 112) placed on either side, one a little before the after turret, and one a little abaft the fore turret; and there are, further, twenty-eight small quick-firing and machine guns, with six torpedo tubes. The protection consists of a 17·7-inch belt of steel at the water-line, tapering to 10 inches at bow and stern, above which is a narrow belt of 3·9-inch armour, which protects the slope and edges of the armoured deck, which is of steel 2·75 inches thick. The armour of the barbettes is 15·7 inches thick, and is carried down at an angle, making a funnelled shape tube to the armoured deck; the barbettes are thus much better supported, and their bases better protected than in the earlier ships, but the protection is still nothing like so complete as it is in our *Royal Sovereign* and *Majestic* classes. The armour on the small turrets for the 5·5-inch guns is 3·9 inches thick. One novelty in these ships is the extensive use of electricity for the barbettes, ammunition hoists, etc., which are all worked by this power.

In the latest type of French battleships, viz., the *Charlemagne, St. Louis,* and *Gaulois* (fig. 113), none of which are likely to be ready for sea for the next two years, the heavy armament, which consists of four 12-inch guns, is mounted in pairs in two closed turrets, one forward and one aft; the secondary armament consists of ten 14 cm. (5·5-inch) quick-firing guns in an armoured central battery, and eight 4-inch quick-firing guns on the superstructure, pro-

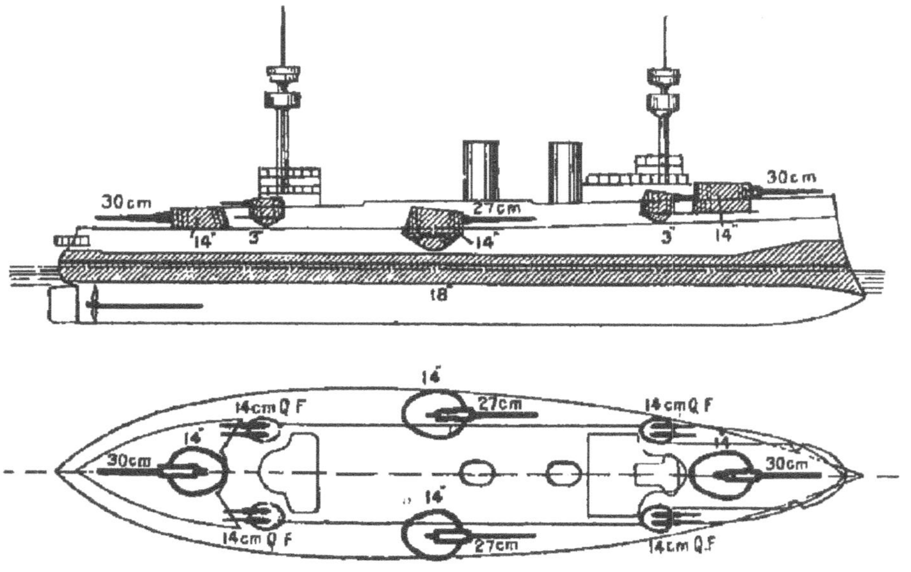

FIG. 112.—PROFILE AND DECK PLANS OF "BOUVET" CLASS, SHOWING DISPOSITION OF ARMOUR AND GUNS.

tected by 2-inch shields. The armour protection is a 15-inch steel water-line belt, tapering to 10 inches at bow and stern, with a narrow 3-inch belt above it; on the turret is 15-inch armour, and the central battery is protected by 3-inch armour. A 6-inch armoured tube connects the turrets with the armoured deck, but the weak spot in these ships is the long extent of unarmoured side between the 3-inch belt and the armoured central battery on the upper deck; moreover, the bases of the turrets are imperfectly protected, and armoured transverse bulkheads, which are found in all our ships, even in the "Admiral" class, are entirely wanting in all the French ships just described.

It may be noticed that the initial, or muzzle, velocity of the French guns is more than ours of a similar calibre; this advantage, and it is an undoubted advantage, has been gained by making the projectile lighter, and by increasing the length of the guns; thus while our new wire 12-inch guns are 40 calibres in length, the French 12-inch guns are 45 calibres; on the other hand, our 12-inch guns throw a projectile 850 lb. in weight, while the projectile from the French gun is only 626 lb. in weight. It is quite certain that our gunnery authorities look with much suspicion upon the great length of the French guns, believing it to be a source of weakness, and have decided to forego the greater velocity in order to insure greater strength in the gun. Like ourselves, the French are discarding their very heavy guns for lighter weapons; the 42 cm. (16·5-inch) 75-ton guns carried by the *Caïman*, and her three sisters are to be replaced in those ships by the new 30 cm. (12-inch) gun, which having a muzzle velocity of 2,600 feet, is a much more powerful weapon than the older, although heavier, gun. It may be interesting to point out that breech-loading guns were first adopted in the French navy in 1860, and no muzzle-loaders have been used since that date.

In the brief description of the chief types of English and French battleships just given, readers will learn the general principles which govern the construction of all such ships in the different navies of the world. Russia, some twelve years ago, laid down three powerful battleships at Nicolaieff and Sebastopol for her Black Sea fleet, viz., the *Catherine II.*, the *Sinope*, and the *Tchesme.* These vessels have an all-round 16-inch water-line belt, tapering to 12 and

10 inches at bow and stern, with a central pear-shaped armoured redoubt protected by 12-inch armour, in which are mounted in pairs, *en barbette*, six 12-inch guns, two pairs in the forepart of the redoubt each side of the ship, and the third in the after part on the centre line; there is in addition a secondary battery of seven 6-inch quick-firing guns, with fourteen smaller quick-firing and machine guns.

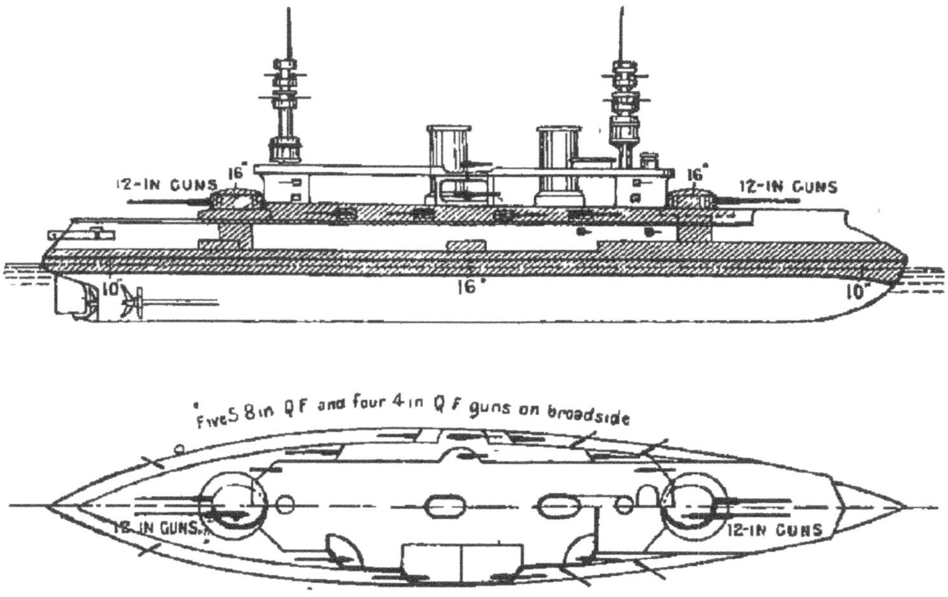

FIG. 113.—PROFILE AND DECK PLANS OF "CHARLEMAGNE" CLASS, SHOWING DISPOSITION OF ARMOUR AND GUNS.

But in all the later ships the Russian Admiralty have adopted a partial belt, as in our ships; with, in the larger ships, the heavy guns mounted in pairs in turrets, one at each end of a central armoured redoubt, and with the secondary armament either coupled in small turrets or in an armoured battery with from 3 to 5-inch armour.

In the German Navy the new four first-class battleships of the *Brandenburg* type, lately completed, are protected by water-line all-round belts of 15·8-inch compound armour, with the guns mounted in pairs in armoured turrets, but in the two newest battleships, now under construction, the water-line belt extends from right forward for four-fifths the length of the ship, and is of 12-inch Harveyized steel, with another belt of 8-inch steel above again, carried up to the upper deck. The four heavy 10-inch guns are mounted in pairs, in turrets, one forward and one aft, while there is a very heavy secondary battery of eighteen 6-inch quick-firing guns, all protected by 6-inch casemates, with, in addition, thirty-two smaller quick-firing guns and machine guns. The peculiarity about these ships is that the size of the heavy guns has been much reduced, 10-inch guns being substituted for the 12-inch guns carried in the latest English and French ships; while, on the other hand, the weight of the secondary armament has been much developed, no other battleships afloat carrying so many 6-inch quick-firers. The new ships are in every way much more formidable vessels than their four predecessors of the *Brandenburg* class, which carry six 11-inch guns, mounted in pairs, in turrets protected by 11·8-inch armour, one forward, one amidships, and one aft, but only have a secondary battery of six 4·1-inch quick-firing guns and these quite unprotected.

The new battleships building for the United States are more or less of the *Royal Sovereign* type, but they carry a somewhat heavier armament with rather heavier plating; but they have a displacement of some 3,000 tons less than the ships of the *Royal Sovereign* and *Majestic* classes.

There is nothing so difficult as to draw true comparisons, even for experts, between different types of battleships and cruisers, especially with regard to their value as fighting machines. Every battleship and cruiser is a compromise. It must always be remembered that only a certain amount of weight can safely be carried on a given displacement;

and if two battleships of the same displacement are compared, and it is found that one carries heavier armour and a heavier armament than the other, it may safely be taken, that to enable this to be done, something else has had to be sacrificed, either in the form of strength of hull, coal-carrying capacity, or ammunition supplies, or very likely in all three. The work that ships have to perform must also be taken into consideration, and in this connection it is as well to point out, that when, as is so often done now by irresponsible writers in England, unfavourable criticisms are passed on our ships, and various foreign types are held up as worthy of imitation, it must be remembered that our battleships are built to keep the sea in all weathers, and that in all our later ships, the coal supply far exceeds that carried in foreign ships of a similar class, and also that our ships carry a much larger ammunition supply than foreign vessels do. Taking them all in all, it may safely be asserted that our battleships of the *Royal Sovereign* and *Majestic* classes are at the present time, from their high offensive and defensive powers, and sea-going qualities, the finest fighting ships afloat.

CHAPTER XVI.

THE FIGHTING ORGANIZATION OF A BATTLESHIP, AND THE MEN WHO FIGHT THE GUNS.

A BOOK of this kind, compiled solely to give non-professional readers some general idea of what naval gunnery is, would not be complete without an attempt now to briefly explain the system on which the crew of a modern war-ship is organized for manning and fighting the guns, and also the methods adopted to make a ship perfect as a fighting unit, fit to perform any duty on which the exigencies of our world-wide empire may require her to be dispatched.

For the purpose of this work it will suffice to give some details of the organization of a first-class battleship—one of the *Royal Sovereign* class, for instance—the whole seven ships of which type being at present in commission either in the Channel or Mediterranean squadrons.

The armament of a ship of this class is made up as follows:—four 13·5-inch 67-ton guns in two barbettes, one forward and one aft; six 6-inch quick-firing guns on the upper deck; four 6-inch quick-firing guns in separate armoured casemates on the main deck; four 6-pounder Hotchkiss quick-firing guns on the upper deck; twelve of the same guns on the main deck; four 3-pounder Hotchkiss quick-firing guns and eight ·45-inch Maxim guns on the shelter deck; and lastly, eight 3-pounder Hotchkiss quick-firing guns in the fighting tops, on the two military masts. The torpedo armament consists of four tubes on the broadside, one stern tube, two submerged tubes, and five search-lights. The following small arms are also supplied:—238 Lee-Metford magazine rifles; 143 pistols and 205 cutlasses, with complete sets of accoutrements and gear for arming the ship's company for active service either on shore or away in the boats of the ship. There are further, in addition, complete sets of armoured cables for submarine or countermines, the mines themselves, and all the necessary electrical plant for running lines of countermines or defending a harbour by observation or mechanical ones.

The magazines and shell-rooms required for stowing the requisite ammunition consist of:—two magazine and shell-rooms for the 13·5-inch guns; two for the 6-inch quick-firing guns; two for the 6 and 3-pounder quick-firing guns; two torpedo magazines, and one magazine for small-arm ammunition; all these are situated in the midship lower part of the ship.

It must be remembered that the question of the ammunition supply of ships, especially since the introduction of quick-firing guns, is a most important one, and that the proportion of ammunition carried by English ships is considerably larger than in any foreign navies. As has been already stated, all war-ships in these days are a matter of compromise, so a large ammunition supply involves sacrifices on points which foreign nations presumably consider of more importance than we do.

The complement of a ship is based on its gunnery and torpedo requirements, and in this case is 712 officers and men, including men specially trained in the use of the war material supplied.

These ratings are:—3 gunnery instructors, usually employed to take charge of the barbettes and for instructional purposes; 14 captains of guns, who are men that have been specially selected as good shots and judges of distance, and who are trained to take charge of the heavy guns; 108 seamen gunners (S. G.'s) and seamen gunner torpedo men (S. G. T.), to take charge of the smaller guns and provide the most important numbers at the heavy guns, torpedo tubes, magazines, and shell-rooms, work the search and incandescent lights, and do all the mining and Whitehead torpedo work; this class also provides seamen divers, and all these men have passed successfully through the gunnery and torpedo schools in this work, while there are in addition 7 leading torpedo men, who take charge of the torpedo tubes and responsible torpedo work.

The remaining numbers comprise seamen and boys who have been trained in gunnery, but not so highly as an S. G., artisans, domestics (these two last being also called day-men, their duties not involving them in regular night-work), marines, and the engine-room staff, the latter being trained in field drills and the use of small arms.

The appropriation or stationing of the ship's company is carried out on the following system:—the men are divided into two watches, "starboard" and "port," each watch consisting of four sections, called after the different parts of the ship, viz., "forecastle-men," "foretop-men," "maintop-men," and "quarterdeck-men," so that if a man belongs, for instance, to the "maintop-men," that is said to be his "part of the ship." Each of these four sections consists of forty-four men, and is again divided into two parts, known as the first or second part of "forecastle-men," etc., as the case may be, and each of these parts into two subdivisions, called the first, second, third, and fourth subdivisions, which each consists of one leading hand and ten men.

Each of the four principal sections is appropriated to its peculiar part of the ship for duty; thus the "forecastle" section of the ship is appropriated to the "forecastle-men," who practically live in their part of the ship, that is, they have their messes in it, clean the upper, main, belt decks and flats attached to the section; man all guns, torpedo tubes, magazines, etc., and supply all ammunition required for that part; so that they have a general and continued interest in that section of the ship.

Thus the men belonging to the forecastle section would be stationed as follows :—fore barbette (11), fore magazine for 13·5-inch guns (20), fore shell-room for 13 5-inch guns (12), fore shelter deck (14), 6-pounders on fore part main deck (12), submerged torpedo tubes (12), ammunition supply (7) ; total, 88.

Foretop-men.—Three 6-inch quick-firing guns port side of upper deck (24), two 6-pounder quick-firing guns same side (6), foretops (15), fore 6 inch magazine (8), fore 6-inch shell-room (5), foremost pair of torpedo tubes (10), fore magazine for 6 and 3-pounder quick-firing guns (4), ammunition supply (16) ; total, 88.

Maintop-men.—Three 6-inch quick-firing guns, starboard side of upper deck (24), two 6-pounder quick-firing guns, same side (6), maintop (9), after 6-inch magazine (8), after 6-inch shell-room (7), after pair of torpedo tubes (10), after magazine, for 6 and 3-pounder quick-firing guns (5), ammunition supply (19) ; total, 88.

Quarterdeck-men.—After barbette (11), after magazine for 13·5-inch guns (20), after shell-room for 13·5-inch guns (12), after shelter deck (14), stern torpedo tube (7), 6-pounders on after part main deck (6), ammunition supply (18) ; total, 88.

Marines.—Four 6-inch quick-firing guns in casemates (32), 6-pounder quick-firing guns on main deck between screens (24), ammunition supply (15), small-arm party and voice tube men (28); total, 99.

Band.—Collect the wounded.

Carpenters.—Fire brigade for hand-worked pumps.

Day-men and Excused Day-men.—Magazines, winches, and assist doctors.

Engine-room Department.—Work at main engines, stokehole, auxiliary engines, close water-tight doors, and provide fire brigade for steam fire main.

In order that the specially trained men should be divided as required, they are placed in the "parts of the ship" as follows:

	Seamen gunners.	Seamen gunner torpedo men.
Forecastle	27	4
Foretop	21	4
Maintop	19	4
Quarterdeck	23	2

The remainder are placed in a special list, which includes quartermasters, boatswains' mates, etc.

In accordance with this scheme, every man on joining the ship is given a station known as his "general quarters" or fighting station, and he repairs to it whenever the bugle sounds "action."

In addition to this, every man who is stationed in a boat has a rifle and sword bayonet appropriated to him, and all coxswains of boats a cutlass and pistol. The remaining small arms are equally distributed amongst the rest of the crew.

The foregoing description will, it is hoped, give a general idea of the organization and stationing of a ship's company for battle.

The system of imparting instruction in gunnery and torpedo work to the crew is for a portion of the guns, etc., to be drilled every forenoon and afternoon, so that the complete system is individually exercised once a week, and these are known as "divisional drills." They lead up to the general exercise which takes place on a Friday forenoon and is called "general quarters," when the complete system is worked at the same time.

An attempt will now be made to take the reader in imagination round a battleship of the *Royal Sovereign* class whilst at "general quarters," and show him some of the usual exercises carried on during the time allotted for that purpose. We will suppose the bugles have sounded "exercise action," and we will start from the conning tower.

In an incredibly short time the electric bell in the conning tower, which is fixed to attract attention, starts, and reports come rapidly in from the different parts of the ship that the guns, torpedo tubes, and magazines are cleared away, and are ready for "action." The order is then given for drill to be carried out by the officers of the quarters, who immediately start every gun and battery on a programme of drills that insures every possible drill and exercise being carried out once a month. On leaving the conning tower we come to the fore shelter deck, and find the 3-pounder and Maxim guns' crews taking out their locks and replacing from spare parts portions that may be shot away in "action," whilst on the after shelter deck the men are having the use of day and night sights explained to them, and are also being exercised in judging distances over the water. The guns' crews in the tops are carrying out aiming drill, *i.e.*, laying their guns for objects whose distance and bearing are constantly altering. Descending to the upper deck, we pass the stoker fire brigade extinguishing a small fire, and come to the starboard side of the upper deck, where the battery is being exercised at the various methods for overcoming a miss fire during electric or percussion firing, and repairing circuits that have been shot away. The port battery is working with diminished crews; *i.e.*, some of them are supposed to have been shot or wounded, and the wounded are being taken away by the stretcher party to the wounded cot station, and are lowered down to the doctor, who is working away in the bath-rooms in the flats. The small-arm men are doing position drill and firing Morris tube at a target, whilst the riggers, under the boatswain, are putting on frapping lines for the rigging, splinter nets about the deck, etc., and the ammunition supply parties are exercising passing up dummies. Leaving this and descending to the main deck by means of a rope ladder, we find the starboard casemates practising independent firing, and the port supplying spare gear that may have been shot away, and tackles for training their guns, supposing the training gear to become disabled. The 6-pounder marine crews are at ammunition instruction and judging distances of objects. We pass on to the belt deck and find all the watertight and armoured doors closed, and a man standing by each one (he also acts as fire brigade in his compartment), and as you pass through the flats you see the doctors at the various hospital depots lecturing to their assistants on wounds, etc., also the carpenter and stokers' fire brigade exercising at small fires and collision quarters, that is, pumping out or draining water from any special compartment, and also the above-water torpedo tubes' crew at drill; you gradually descend to the ammunition passage that leads into all the magazines and shell-rooms, and find the men here exercising at flooding the magazines, and supplying ammunition under every practical combination of the supply from the usual magazines being cut off. Ascending into the barbettes we find the after barbette drilling with hand-loading gear, under the supposition that the hydraulic gear has broken down, and the fore barbette is exercising with relief crews, that is, crews formed from their magazine and shell-room parties.

It is now time to go back to the upper deck, when the "cease firing" is sounded, and a second series of drills commence. The first bugle is "prepare to ram," and all the guns are trained on the bow, and laid ready to pour a broadside fire into the ship that is being rammed, every man not actually required to fire the quick-firing guns takes cover, so as to protect himself from the enemy's fire; a broadside is fired and all the guns are brought back into the cease firing position after the opposite broadside has trained rapidly on the quarter and delivered its fire at a ship that has been missed by the ram. The next general exercise is boarders, and one after the other each of the three divisions of boarders is brought up and exercised at boarding and repelling boarders in different parts of the ship; these divisions are formed by the different quarters, and are called up from the side of the ship that is not engaging the enemy. On returning to their quarters, the order is given "upper deck quarters, out collision mat," and the mat is got out and placed over a supposed wound in the ship's side, and the pumps are got ready for pumping out the compartment in which the injury has occurred. The guns are then secured, ammunition returned, magazines and shell-rooms closed. Both watches are ordered to fall in for exercise, and each watch is exercised at "man and arm ship" to resist a torpedo attack; all the light quick-firing and two upper deck 6-inch guns are cleared away and loaded; all the guns manned are divided into four sections, and they only fire over their allotted areas, that is from right ahead and right astern to abeam, on each side of the ship; search-lights and a large number of special look-out men, in every part of the ship, are employed to detect the presence of an enemy. Each section is exercised by steam siren or bugle, and when the "cease firing" and "disperse" sounds the "general quarters" drill is over. This goes on each Friday, and on each occasion during the month the drills are varied.

In additiou to the drills the ship goes to sea each month and expends a portion of her quarterly allowance of ammunition, steaming past a target at a high rate of speed, and also makes runs with her torpedoes from all the tubes, while every month, or more frequently if possible, a large number of rounds are fired from cannon or aiming-tubes, inserted in every nature of guns; they allow of 1-inch, ·45-inch and Morris tube ammunition being expended from these guns under similar sighting conditions as if the guns were loaded with their proper ammunition. Whilst in harbour all the principal numbers at the guns fire from a 3-pounder aiming-tube from a torpedo boat steaming rapidly past targets, and the ship's company generally rig countermine launches, lay out mines, run Whitehead torpedoes from ship and boat, practice creeping and sweeping for submarine mines, firing outrigger charges, fire away ammunition from guns in boats, land field and Maxim guns for firing from the shore; and, in conclusion, there is not an exercise that the ship's company is not put through every quarter, and most of them once a month, including "night quarters" or clearing away the guns for action in the middle of the night.

In preparing the ship for battle, everything that masks the fire of the guns is removed, such as davits, stanchions, etc.; all boats and everything that is liable to catch fire is got rid of, wooden ladders are removed and rope ladders fitted in their place, splinter nets got into position all over the ship, coffer-dams filled, extra protection with chain cables, spare boiler-plates and coal bags arranged for upper deck quarters, depôts of ammunition formed in protected places in different parts of the ship, and a drinking-water supply arranged along each deck.

All the foregoing exercises and drills have to be arranged so as to work smoothly and in conjunction with each other, until, after a ship has been a short time in commission, the perfect fighting machine develops from this intricate mass of war material, and is ready to repel any attack that may be made against her, or to deliver a crushing blow to an antagonist, as it only requires the pressure of a single electric button, or a bugle call, to let loose, with terrific force, the enormous pent-up powers that are lying ready in her guns, the simultaneous discharge of which will make the biggest battleship, monster though she is, shiver and heel to the shock of their discharge.

In addition to their duties on board, bluejackets are often called upon to operate against an enemy on shore, and for this purpose they are organized as a landing party. The battalion for landing consists of four companies, formed from men in the forecastle, foretop-men, maintop-men, and quarterdeck-men, armed with rifles, and are known as the A, B, C, or D companies. The rifles issued to the ship are equally distributed amongst the four parts, so that about sixty men are in each company.

Two 9-pounders, R. M. L., mounted on field carriages with limbers, and two Maxims on travelling carriages, are supplied for use with this force, and each gun is attached to one of the companies, the numbers, as far as 4, who work them being selected from coxswains of boats and men armed with pistols, and the remainder (14 for 9-pounders and 8 for Maxims) are men armed with rifles taken from the companies to which the guns are attached. Each company has a signalman, armourer, pioneer, stretcher men (4), water carriers (4), spare ammunition men (8), men with spare parts for rifles and cleaning gear (2), markers, and petty officers attached to them, also two midshipmen and one lieutenant in command. Each gun's crew has a similar extra party with it. To accompany the whole battalion generally, there is a telegraph party, with wire and instruments (12), gun-cotton torpedo party (8), search-light party (8), hospital party with doctors (4), commissariat party, including cook and mate (10), baggage party (8), officers' servants (6), orderlies (4), buglers (4), band, and the usual battalion officers.

Each man has a complete set of accoutrements, which includes ammunition pouches, bandoliers, water bottle, haversack, blanket, and all the front rank Wallace intrenching spades.

When drawn up on shore they form a small but complete fighting force in every particular, with an artillery and machine-gun battery attached to it and all the appliances for marching and taking the field for any lengthened period. They are taught and are able to carry out all regimental manœuvres of attack and defence; they are annually put through a course of musketry practice, and carry out Morris aiming-tube practice with rifles weekly.

It is sometimes necessary to send away boats' crews "manned and armed" for river and cutting-out work. For this purpose a "man and arm boats" organization exists. Each large boat has a 3-pounder quick-firing gun and a Maxim gun mounted in it, and the cutters a Maxim, so that a ship of the *Royal Sovereign* class sends away in her boats four 3-pounder quick-firing guns and seven Maxims, besides the boats' crews, rifles, pistols, and cutlasses. The other stores taken away in the boats include ammunition, water, provisions, spare gear for guns, signal lights and rockets, whilst each large boat carries an armourer, signalman, and carpenter, and all boats extra men for working the guns; surgical bags for the wounded, boatswains, boats, and carpenters' bags, containing the gear for small repairs and boat work, marlingspikes, lead and lines, etc., completing the fit-out.

The boats of a ship are frequently "manned and armed" for exercise, and are completed for this service in under five minutes from the time the order is given.

The reader can now form some idea of the various duties which a bluejacket has to perform, and will realize that he has to acquire a varied and extensive knowledge far superior to that possessed by the class from which he is drawn, and it is this fact that makes him such a valuable unit in the defence of the British Empire. But as unfortunately there are a very large number of educated, and even on other subjects well-informed, people in England to whom the term bluejacket conveys no distinct meaning, and who are unaware of the difference which exists between the smart, well-educated and highly-disciplined men, who now man our ships of war and the ill-kempt, slouching, in many cases quite uneducated, sailors, who are to be found forming the crews in only too many of our merchant vessels, it may be of use to briefly sketch how our bluejackets are entered and trained to fit them for the various services they are called upon to perform.

Unlike the Army, where a system of short service prevails, long service is the order of the day in the Navy, and without it our men could not be trained to be what they are. An ignorant, slouching lout can be taken from the plough-tail for the Army, and under the hands of a smart drill-sergeant will, in a few months, be turned out a smart, well set-up soldier, an efficient fighting machine, proud of himself and his calling, but in the great majority of cases he is simply a fighting machine and nothing else. Now in the Navy a good deal more is required than that. A bluejacket must be a good fighting machine, it is true, but he must be other things as well. Above all, he must have a head on his shoulders, and be able to act and think on his own account; and to become what is termed a first-class seaman-gunner and torpedo-man, he must, for his position in life, be remarkably well educated, and it is also evident, that his thorough training, instead of being, as in the case of a soldier, a matter of months, must be, what it is, a matter of years.

Our bluejackets, or "seamen proper," as distinct from the artificers, stokers and marines, are entered as boys between the ages of 15 and $16\frac{1}{2}$; the number entered yearly in the different training-ships being now close on 6,000. In view of the lamentable want of knowledge displayed by many writers in the Press at the present time, with regard to what they call the lack of men in the Navy and the supposed difficulty of obtaining men for the Service, it may be as well to state that since the introduction of the present system of long service (now some thirty years old) and of entering our seamen as boys, no difficulty has ever been experienced in obtaining as many boys of the best class as the Admiralty have decided to enter each year; nay more, so popular is the Service and so large the number of eligible candidates, that twice during the last eighteen months (middle of 1895-end of 1896) the Admiralty have been able to raise the standard (in height and chest measurement), and nearly 40 per cent. of the candidates each year are rejected. So much for the difficulty of getting boys and the unpopularity of the Service. The question of whether we are short of men or not merely depends upon whether in any given year or number of years a sufficient

West & Co., Southsea, photo. *To face p.* 340.

CUTLASS DRILL ON QUARTER-DECK OF "ROYAL SOVEREIGN."

"SLOPE SWORDS."

number of boys have been voted for by Parliament, and not because there has ever been any difficulty in getting boys to join.

To show how carefully the boys to form our future blue-jackets are selected, it may be as well to mention that a candidate has to be able to read and write well, and that the superintendent of police for the district where he lives has to furnish a confidential report, not only as to the good character of the boy himself, but as to his parents, and if this character is not satisfactory the boy is rejected. When such pains are taken to secure superior lads for the Service, and when under such stringent conditions as to qualifications for entry, we can still pick and choose and yet find no difficulty in procuring the right sort of lads, it is not to be wondered at that the Admiralty steadily set their face against entering boys from industrial and reformatory ships, in spite of strong pressure brought often to bear upon them to do so. With regard to boys from reformatory ships, it would be obviously most unfair to the parents of the boys, not to mention the boys themselves entered under the above strict conditions, that they should be subject to the risk of contamination from lads brought from reformatories, while the boys from the industrial ships, as a rule, are wanting in physique, and so fail to pass the strict medical ordeal through which alone they can enter the Service. A good many boys, however, are entered from the mercantile training-ships *Arethusa, Indefatigable, Warspite, Exmouth*, and *Mercury*, where the boys are all carefully selected for entry, none of bad character being received on board.

When accepted for entry in the training-ship the parents or guardians of a boy have to sign an agreement undertaking on his behalf that he will serve for twelve continuous years from the age of eighteen, and he is then rated a boy, second class, and after one year's service, if he comes up to the requisite standard in seamanship and rifle and cutlass drill he is rated a boy first class. Boys remain in the training-ships from one to two years, according to their age and proficiency; a part of this time being spent on board one of the brigs attached to the different training ships. There are seven of these boys' training-ships (exclusive of their tenders) stationed round the coast: the *St. Vincent* at Portsmouth, the *Boscawen* at Portland, the *Impregnable* (the headquarters' ship of the Training Service) and the *Lion* at Devonport, the *Ganges* at Falmouth, the *Black Prince* at Queenstown, and the *Caledonia* at Queensferry, N.B. Part of the time in the training-ships is spent on board the brigs attached to each stationary training vessel; and after the boys become ordinary seamen they are, as far as possible, made to go through a further period of education in practical seamanship on board the Training Squadron, or the small sailing training cruiser attached to the Mediterranean Fleet.

At the age of eighteen, or at latest eighteen and six months, a boy is rated ordinary seaman, and it then depends upon himself how soon he obtains his next step, which is that of able seaman. A smart lad will probably be rated A.B., as it is called, by the time he is twenty or soon after, but before obtaining the rating he must pass, besides the necessary seamanship examination, a qualifying examination in gunnery, becoming what is called a T.M. or trained man; the passing for T.M. gives him an extra penny a day, whether as an ordinary or able seaman; his pay as an ordinary seaman amounts to 1*s.* 3*d.* a day, and as an A.B. to 1*s.* 7*d.* a day. As an A.B. he will in due course go through the gunnery and torpedo schools, and if smart, and he has not neglected his school opportunities, since he left the training-ship, he may obtain his certificate as a seaman-gunner of either the first or second-class; if first-class, he gets an addition to his pay of *4d.* a day, if second-class of *2d.* a day. If he obtains a first-class certificate as seaman-gunner, he may go through the torpedo-school, and on obtaining the requisite certificate become an S.G.T. (seaman-gunner torpedo-man), when he will receive *6d.* a day additional pay. All men, whether petty officers or otherwise, who re-engage for a second term of ten or twelve years' service, receive an additional *2d.* a day pay, and become entitled to a pension at the end of their term of service. From A.B. a man may rise to leading seaman, when he receives 1*s.* *9d.* a day, his next step being to second-class petty officer with 2*s.* a day pay; he can then become a first-class petty officer with 2*s.* *2d.* a day, increased to 2*s.* *5d.* a day after four years' service in that rank. Should he then be promoted to chief petty officer, he receives first 2*s.* *8d.* a day, and after three years' service as such 3*s.* *2d.* In all the above rates of pay, the additional *2d.* for second period of service has not been included. There are further additions a man can make to his pay by qualifying for and being appointed to perform certain superior gunnery duties; for instance, as a gunnery instructor he will receive *8d.* a day in addition to his *6d.* a day as a first-class seaman-gunner and torpedo-man, a total increase of 1*s.* *2d.* to his ordinary pay; as a captain of a turret he receives *3d.* a day, and as a captain of a gun, *2d.* a day in addition to his other gunnery allowances, while every man receives an extra penny a day for each good-conduct badge he has up to three. A man can obtain his first badge after three years' service from the time he obtains his rating as ordinary seaman, his second after eight years' service, and his third after thirteen years' service. It is therefore possible for a chief petty officer to be receiving as much as 4*s.* *9d.* a day, including his gunnery allowances and good-conduct badge pay. After ten years of very good character a man becomes entitled to the good-conduct medal, and after fifteen years of such service to an additional gratuity on receiving his pension, to which men are entitled after twenty years' service in man's rating, and if of good character. Men can re-engage for a third period of service, and of course the longer they serve the better their pensions are on finally quitting the service.

A man may become a candidate for promotion to the rank of warrant officer any time between the ages of twenty-one and thirty-five on passing the necessary examinations, but before receiving his promotion he must be a petty officer with not less than one year's service in that rating and must further have been seven years at sea, during which time his character must have been such as would, if continued, qualify him for the good-conduct medal. The pay of a warrant officer begins at 5*s.* 6*d.* a day, rising gradually to 9*s.* after fifteen years' service, and there are certain allowances, varying from 6*d.* to 2*s.* a day according to the duties performed. A warrant officer can rise to chief warrant officer, when he receives 10*s.* a day, rising to 12*s.* after eight years' service. A certain number of chief warrant officers are now granted the honorary rank of lieutenant when they retire.

The foregoing description will give a good general idea of the condition of service under which our bluejackets are trained, and the nature of the career open to them. In no other navy in the world are the men subjected to so long and careful a course of training as in ours, and by the time a man has become a first-class seaman-gunner and torpedoman, he is of considerable value to the country and well worth all the money spent on his education; able to perform his duty equally on land as at sea, equally at home in the barbette of a battleship or skirmishing on shore, steering a torpedo boat or working a field-gun or rocket-tube. Although there has been no great naval war in which we have been engaged since 1815, yet during the last eighty years the Navy has by no means led a life of inglorious peace. In 1816 Algiers was bombarded and the power of the Dey broken; in 1824 came the first war with Burmah, carried out successfully by the combined naval and military forces; in 1827 the battle of Navarino was fought; in 1839 began the first war with China, in which the Navy took its full share, while 1840 saw also the operations on the coast of Syria and the bombardment of Acre; in 1851-53 the Navy was again engaged in the second Burmese war; in 1854 came the war with Russia, where, in addition to the operations in the Baltic, Black Sea, and Sea of Azov, 3,000 men were landed from the Fleet to man the batteries; in 1856 the second China War began, not finally concluded until 1860; during the Indian Mutiny one Naval Brigade accompanied Lord Clyde's force, while another served through the whole campaign under Sir Hugh Rose; in fact, since the Russian War there has been no single campaign, except the Afghan and India frontier wars, in which the Navy has not taken a part both on its proper element the sea, and in the operations on shore; in this connection it may be mentioned that in the Abyssinian campaign the only body of men in the whole army which arrived before Magdala after the toilsome march of 400 miles over the mountains, without one single man having fallen out through foot-soreness or other cause, was the Naval Brigade. I have drawn attention to these facts, as it shows the necessity which exists in these days that our bluejackets should have, in addition to a thorough knowledge of gunnery in all its branches, and of their duties as seamen, a sufficient amount of military training to enable them to act efficiently on shore, either with field-guns, rockets, or as infantry, when, as is now almost always the case, they are called upon to take their part in the numerous operations which we are so continually obliged to carry out in all quarters of the globe.

CHAPTER XVII.

CONCLUDING REMARKS.

ALTHOUGH the battle between guns and armour is still being fought out, it seems probable that the gun, as in the past, so in the futnre, will continue to have the best of it. The amount of armour a ship can carry is limited, while, on the other hand, in the new guns, aided by more powerful explosives, the velocity of the projectiles and consequently their penetrating power, have been and are still being much increased. One of the disadvantages, however, attaching to the modern breech-loading guns is their great length, which is not only a source of weakness in itself, but renders them much more vulnerable in action than the older weapons. The late Mr. Longridge, to whom, as has been already mentioned, our new system of wire guns is due, considered that a still further increase of ballistic power could be obtained for our guns. He was of opinion that in cordite and other smokeless powders now in use, we had propellents so far superior to the powder of only a few years ago, it was unlikely that there would be any much greater improvement effected in that direction. He looked, however, to a further development of ballistic power, (1) by an alteration in the gun itself; (2) by an increase in the efficiency of the present powders by adopting a higher maximnm pressure. He held that guns of extreme length do not give increase of power in anything like the proportion of their length, and taking into consideration their unhandiness and liability to damage, due to their excessive length, that no improvement of ballistic power could be looked for by any further lengthening of the guns, and that an increase in such power could only be looked for by altering the conditions under which the powder is used. With cordite the maximum pressure fired in a close vessel with gravimetric density = 1, is about 170 to 190 tons per square inch; whilst as now generally used in guns, it is limited to about 15 to 20 tons per square inch. The old powders, of which the maximum pressure in a close vessel did not exceed 43 tons per square inch, were used with about the same maximum pressure of 17 tons, so that, as regards the maximum pressure in use, no advantage has as yet been obtained from the new powders, as it is much the same at the present day as was used thirty or forty years ago; the main reason for this being that until the last few years guns would not safely resist a greater pressure. Mr. Longridge, however, maintained that since the introduction of wire guns, this reason no longer has any force, as wire guns may be made to resist an internal pressure of 30 or 40 tons per square inch, with as great a margin of safety as the forged steel gun previously in use under a pressure of 17 tons per square inch, and with far greater security as regards defects of material and workmanship. Taking the case of the 12-inch Woolwich wire guns, firing a projectile of 850 lb. with a charge of 148 lb. of cordite, which give the following ballistic result:

Maximum pressure . .	16·5 tons per sq. inch.
Muzzle velocity . . .	2,328 feet per second.
Muzzle energy . . .	31,800 foot-tons.
Penetration of wrought iron .	25·74 inches.

Mr. Longridge held, that using the same projectile and charge, but with a reduced powder chamber, so as to give a maximum pressure of 27 tons per square inch, the result would be:

Muzzle velocity . . .	2,694 feet per second;
Muzzle energy . . .	42,741 foot-tons;
Penetration of wrought iron .	29·85 inches;

showing an increase of energy and penetration power of about 34 per cent.

If, again, with the same gun and projectile, the pressure were raised to 27 tons per square inch by increasing the charge to 207 lb. of cordite, the result would be:

Muzzle velocity . . .	2,885 feet per second;
Muzzle energy . . .	49,032 foot-tons;
Penetration of wrought iron .	31·98 inches;

showing an increase of 54 per cent. in energy and penetration, which shows that the present guns may be greatly increased in power by adopting a pressure of 27 tons instead of 16·5 tons per square inch. Mr. Longridge was further of opinion that equally good results as those now obtained could be had from guns much reduced in length, provided the pressure is increased; and instancing again the 46-ton 12-inch wire gun, he maintained that the length of travel of the projectile, which is now 30 feet, could be reduced to 14 feet 6 inches, and that with a pressure of 27 tons the ballistic result will be the same as it is in the present guns with 16·5 tons per square inch. The gun in this case would be about 24 feet long over all instead of 37 feet, and would weigh about 35 tons instead of 46 tons as at present. Other smaller wire guns could similarly be reduced in length, if the pressure is increased. Mr. Longridge's system of wire guns has proved so successful, that probably, before long, we shall see experiments carried out with a view of testing his latest theories, and should they be confirmed, it would enable us to add immensely both to the offensive and defensive powers of our ships.

Great as has been the advance in guns, it cannot be denied but that an equally great advance has been made in armour, and there is little in common between the nickel and chrome steel armour plates of to-day and the iron plates on our early battleships. Wrought iron was used by us as armour up to as late as 1880, when the *Inflexible* was completed, but the introduction of breech-loading guns, with their vastly increased penetrative powers, had shown by this time that something more for defensive purposes was required than was afforded by wrought iron, as it was impossible to increase indefinitely the thickness of the plating. The French had before this adopted steel for their armour, and in the handling and manufacture of large masses of steel their manufacturers were far in advance of ours. There were, however, many objections raised in England at that time to the adoption of steel alone for armour. Experiment showed that while up to a certain point it was extremely difficult for a projectile to penetrate a steel plate, in consequence of its hard surface, yet, when once penetrated, the plate was liable to break up entirely, which was not the case with wrought-iron plates. As the result of long experiment the Admiralty finally adopted what is known as compound armour, that is, iron faced with steel, it being found that while the steel face broke up many projectiles, if the projectile did penetrate, the wrought iron rear portion of the plate tended to hold the whole plate together, instead of breaking up completely, as so many of the steel plates did when they were penetrated. So successful did some English firms, notably Cammel and Brown, of Sheffield, become in the manufacture of this species of armour, that all the heavy armour-plating, even in the *Royal Sovereign* class, is compound, and Russia and Italy also adopted this system for their ships; France, Germany, and Austria, however, kept to solid steel. Compound armour, however, is now practically obsolete, and it is to the United States that the introduction of the hard-faced armour universally coming into use is due. What is known as the Harvey hardening process, or some process analogous to it, has now been generally adopted by the English and Continental armour-plate manufacturers, its characteristic features being the application of carbon at a very high temperature for a long time, followed by water hardening. It was for some time believed that the Harvey process, however successful it proved with armour plates of a medium thickness, was not likely to be so successful when applied to thick armour, but 18-inch Harvey steel plates, manufactured by the Carnegie firm in the United States for one of the new battleships, appear to have stood the ordeal, when attacked by 12-inch projectiles, weighing 850 lb., with great success. The Carnegie Company have also adopted a system of "double-forging" plates, which have shown remarkable powers of resistance. Nickel is now extensively used with steel plates, and very remarkable results have been obtained by Messrs. Krupp, the great German firm, with nickel steel hardened plates of various thicknesses, including some 11·81 inches thick, which broke up all the projectiles fired against them. Harvey armour is used in our battleships of the *Majestic* class, while nickel steel Harvey plates are to be used for the new ships of the *Canopus* type, the nickel appearing to render the carbon more sensitive to hardening, so that Harvey plates of nickel steel are toughened at depths hardly affected in simple steel plates. In conclusion the following list of the principal armour-plate and gun manufacturers may be of interest. In England the three great armour plate firms are those of

Messrs. Cammel, Brown and Vickers of Sheffield, but new firms, notably Messrs. Beadmore of Glasgow are now successfully competing for nickel steel armour; in France, Messrs. Schneider and the St. Chamond, Chatillon-Commentry, Marrel Frères and St. Etienne Companies; in Germany the firms of Krupp and Dillinger; in Austria the Witkovitz Company, all of which are remarkably successful in their production of armour plates; in Italy all the armour plates for the fleet are made at the large steel works of Terni; while in the United States the Carnegie and Bethlehem Iron Works Companies both compete successfully with their European rivals. In England the bulk of our guns are made at Woolwich in the government arsenal, but a certain number come from Elswick, while the mountings are manufactured mostly by Elswick and Messrs. Whitworth and Co. Lately the Elswick and Whitworth firms have amalgamated, as have also Messrs. Vickers and the Naval Armament Construction Company of Barrow-in-Furness, while in France it is reported that the Canet and Schneider firms have done the same. In France the larger proportion of the guns for the fleet are manufactured at the government factories at Ruelle, but the Canet firm also furnishes some. In Germany the Krupp firm at Essen provides all the guns for the government, as it does for many foreign governments. In Italy the guns are now manufactured almost entirely at the Armstrong manufactories which were established a few years ago at Pozzuoli, near Naples. Austria obtains her guns from Krupp, except the quick-firing ones, which are now manufactured at the Skoda works at Pilsen in Bohemia; while in Russia the heavy guns, including 6-inch quick-firing guns, are made at the Obukoff gun manufactory; the smaller types of quick-firing guns are made at both the Obukoff and Toulla gun works, the gun carriages and mountings being manufactured at the Obukoff factory and the Newsky steel works.

INDEX.

"A" Magazine case, 191.
Accidents, *Collingwood*, 93; *Duilio*, 65; *Thunderer*, 64.
Agamemnon, 77.
Ajax, 77.
Alcide, 17.
Alexandra, 302, 303.
Amiantive, 22.
Amiral Baudin, 318.
Ammunition, 151, 163, 188, 189, 192, 200, 215.
Andrea Doria, 121.
Annealing, 58, 99.
Anson, 119, 308, 309.
Arbuthnot, Lieutenant, 203.
Ariadne, 18.
Armour, 44, 47, 79, 294; penetration of, 44; piercing projectiles, 247.
Armstrong guns, 35-43, 45, 241.
—— Lord, 35, 41, 43, 45, 80.
—— projectiles, 40, 243, 244, 245.
Artillery, early, 1-9.
Attachment of shields, 316, 317.
Augmenting strips, 124, 243.
Automatic sights, 203.
Auxiliary elevating gear, 157.

Baessys or bases, 5, 7.
Bags, burster, 250; cartridge, 21, 221, 226.
Ballistic power, 347, 348, 349.
—— results, 83, 348, 349.
Barbettes, 308.
—— or Turrets, 119, 133, 153.
—— —— *v.* Broadside mountings, 299, 300.
Barfleur, 113, 117, 313
Base fuzes, 284, 285, 286.
Basilisk, 6.
Bastard cannon, 6, 10; culverin, 6.
Batteries, electric, 180, 205, 206; secondary, 113, 174, 307, 312, 320, 325, 326.
Bays, magazine, and shell-room, 191, 193.
Bellerophon, 114, 301.
Benbow, 121, 123, 308, 309.
Bentham, Sir Samuel, 16
Bernard de la Tour, 1.
Bethlehem Ironworks, 352.
Between-deck mountings, 142, 185-187
Breech-loading *v.* Muzzle-loading, 49, 79-83.
Black Prince, 295, 342.
Blake, 314
Blenheim, 314.
Bogies, 193.

Bollards, hydraulic, 192.
Boring guns, fine, 62, 103 ; rough, 58, 99.
Bouvet, 322, 323.
Boxer fuze, 268, 277 ; rocket, 240, 261, 262 ; shrapnel shell, 236.
Brackets for shields, 317.
Brandenburg, 325, 326
Breaking strength, 57, 99.
Breech, closing, 7, 8, 33-36, 84 ; loaders, 7, 8, 32, 33-36, 79.
—— mechanism, 38, 39, 40, 84-87, 126-130, 177-182 ; Hotchkiss, 165-168 ; hydraulic, 130-133 ; single-motion, 187.
—— screw, 39, 84-85, 92, 126-133 ; coned, 177, 178.
Bridges of electric tubes, 293.
Broadsides, 29, 30, 209.
Broadside mountings, 12, 24, 26, 69, 74, 142, 185.
Brown, 350, 352.
Buffers, hydraulic, 74, 75, 76, 137-141, 153, 154, 168, 169, 177 ; slide, 130.
Building up guns, 35, 59, 100-103
Burster bags, 250, 254.
Bursters, 236, 237, 245, 250, 254.
Bursting of guns, 64, 65, 93.
Bye-pass valve, 140, 141, 144.

Cæsar, 314-316.
Cage, 193, 198, 199.
Caiman, 318
Caledonia, 342
Cammel, 350, 352.
Camperdown, 119, 120, 308, 309.
Canet, 352.
Cannelures in driving-bands, 56, 124, 251, 252.
Cannon, first use of, in ships, 1 ; demi, 5, 6, 9, 11 ; royal, 6 ; petro, perers, or periors, 5, 6, 7, 8 ; serpentine, 6.
Canopus, 125.
Captain, 303.
Carbon, 96, 97, 99.
Carnegie, 351, 352.
Carnot, 322, 323.
Carriages, gun, 4, 12, 24, 26, 66-76.
Carronades, 14, 15.
Cartridge cases, 11, 28, 179, 190 ; quick-firing guns, 178, 228.
Cartridges, 5, 13, 62, 81, 179, 190, 220, 221.
Casemates, 312, 314, 316, 326.
Case-shot, 232, 244, 255
Cases, powder, 190, 191, 222.
Catherine II., 324, 325.
Cavalli, Major, 33, 34.
Central loading, 196, 197, 199.
Centre-pivot mountings, 144-148.
Centurion, 117, 149, 199, 313.
Chambering, 81, 82.
Chambers, 8, 32, 33, 66, 81, 82, 122.
—— quick-firing guns, 174.
Charlemagne, 323, 324, 325.
Charles Martel, 322, 323.
Chase-hooping, 94.
Chen Yuen, 257.
Clip-plate, 142, 148 ; ring, 142, 144.
Coles, Captain Cowper, 298, 299, 300, 303.
Collingwood, 93, 118, 308.
Colossus, 118, 307.
Commission, *Thunderer's* accident, 64.

Committee on H.M.S. *Collingwood*, 93.
—— Armstrong and Whitworth guns, 45.
—— Ordnance, 89, 108.
Common shell, 246, 249.
Compressors, 68, 69, 74, 75.
Coned breech-screw, 177, 178.
Congress, 297.
Congreve, 260, 261.
Conqueror, 118.
Construction of guns, 35, 37, 52, 57-62, 96-111.
Controlling rams, 146.
Conversion of smooth-bore guns, 50.
Cordelière, 3.
Cordite, 223-225.
Corrugated cases, 190, 191.
Cradles, quick-firing, 177, 182, 183, 187.
Crakys of War, 1.
Crimea, 34, 35.
Culverin, 2, 6, 9, 10; bastard, 6.
Cumberland, 297.
Curtall, or curtow, 2.
Cylindrical magazine cases, 190, 191.

Dandolo, 57.
De Bange pad, 86, 87, 88, 115, 130.
Demi-cannon, 5, 6, 9, 10, 11; culverin, 5, 6, 9, 10; slings, 5.
Dermatine, 190.
Descharges, 2, 3.
Devastation, 117, 304, 305.
Diadem, 18, 314.
Direct-action fuzes, 281, 282; Delay fuzes, 282.
Dispart sights, 22, 24.
Doris, 18.
Douglas, Sir Charles, 19, 21
Douglas, Sir Howard, 21.
Driving-bands, 124, 251.
Drying, 218.
Duke, 19.
Duilio, 57, 65.
Dusting, 218.

Early cannon, 1-8, 32
Edgar, 314.
Edinburgh, 118.
Elasticity, limit of, 57, 91.
Electric firing gear, 179-182, 205; primer, 179, 206; tubes, 205.
Electric *v.* percussion firing, 179-182.
Elevating gear, 73, 141, 142, 145; hydraulic mountings, 77, 153, 156.
Elongated projectiles, 40, 47, 48, 49, 241-257.
Elswick, 43, 65, 69-71, 74, 79, 80, 83, 84, 86, 89, 102, 121, 145, 168, 172, 175, 177, 186, 187, 203, 208, 352; cup, 86, 87.
Empress of India, 119, 314-316.
Endurance of guns, 123.
Erosion, 56, 105, 124.
Essen, 352.
Euryalus, 41.
Experiments, 33, 34, 35, 45, 80, 81.
Extractor, quick-firing guns, 168, 178, 179.

Falconet, 6.
Falcon or fawcon, 5, 6.
Firing gear, 85; quick-firing guns, 179-182; keys, 205; stations, 157, 204, 213, 214.
Foote, Lieutenant, 203.

Fore sight, 202, 204.
Formidable, 19, 319.
Fossam powder, 66.
Fowler, 5, 7.
Francesco Morosini, 121.
Fraser construction, 59, 64.
Friction tubes, 292.
Fuzes, 263, 293; base, 284, 285, 286, 287; Armstrong, 285, 286, 287; Hotchkiss base, 284, 285; Nordenfelt, 285; pillar, Armstrong, 277.
— percussion, 274; Boxer, 277; direct-action, 281, 282; direct-action delay, 282; Freeburn, 274; Moorsom, 275; Pettman, G. S, 277, 278, 279; Plain, B. L, 280; R. L., 280, 281; small, 282, 283, 284.
— time, 266, 267; Armstrong, E., 271, 272; Boxer, 268, 269, 270, 271; Boxer, with detonator, 270, 271; mortar, 267, 268; and concussion, 287, 288; and percussion, middle and short, 289, 290; sensitive middle, 272, 273, 274

Galatea, 18.
Ganges, 342.
Gardner gun, 159, 160.
Gas checks, 56.
Gatling gun, 159.
Gaulois, 323-325.
Girder strength, 101, 107.
Glazing, 218, 219.
Granulating, 219.
Grape shot, 282, 283.
Gravière, Vice-Admiral De la, 17.
Grenades, hand, 17.
Green charge, 218.
Grenfell, Captain, 207, 208.
Grooves, 33, 36, 40, 49, 51, 52, 53, 54, 55, 103, 113, 122.
Gun carriages, 12, 24, 26, 66-78; construction, 35-40, 57-64, 89-95, 96-111; cotton, 224, 225; metal, 135; mountings, hand-worked, 66-78, 133-151, 160-186; mountings, hydraulic, 152-158; steel, 57, 89, 90, 91, 96-102.
Gunpowder, 79, 80, 81, 215-224; prismatic, 219-222, 224; smokeless, English, 223-227; French, 228, stowage of, 188, 192.
Guns, B. L, 2, 4, 7, 8, 32; B. L R., 33, 35-46, 79-134; M. L. R., 47-65; Q. F., 159-187; Hotchkiss, 164-170; Maxim, 160-163; Nordenfelt, 159, 164, 165, 170; wire, 106, 111, 125, 126, 316, 347-349.

Half-charges, 220.
Handing-rooms, 189, 190.
Hannibal, 314-316.
Hardening, 58, 90, 91, 99, 100.
Harvey process, 351.
Henry Grace-à-Dieu, 4, 5, 7, 13.
Hero, 118.
Hidden's patent lock, 20, 21.
High explosives, 256.
Hoche, 320.
Holtzer projectiles, 249.
Hood, 119, 311, 312.
Hoops, 90, 93, 94, 108, 113-126.
Hotchkiss, base fuze, 284; guns, 164-170.
Housing, 185, 186, 187.
Hydraulic gun-mountings, 77, 137, 149, 151, 152, 153, 158; recoil presses, 75, 76, 137-141, 145-147, 154, 168, 169, 170.

Hydraulics as power, 151, 152.

Illustrious, 314-316.
Impérieuse, 200.
Impregnable, 342.
Increasing rifling, 53, 54.
Indomptable, 318.
Inflexible, 57, 78.
Ingot heads, 98.
Ingots, steel, 57, 58, 89, 96, 97, 98.
Inside bolt, 134.
Internal strains, 58, 97, 99, 100.
Iron ores, 96

Jacket recoil mounting, 165, 177, 182.
Jackets, 60, 113, 114, 117, 118, 120, 125, 126.
Jauréguiberry, 322, 323.
Joints, bayonet, 92 , hook, 60.
Jupiter, 314-316.

Key valve, 139.
Krupp, 84, 352.

Lacquering shell, 245, 248, 254.
Ladles, casting, 97; powder, 13.
Lancaster guns, 34, 35.
Length of guns, 80, 81, 82, 347, 348, 349.
Life of guns, 123.
Lifts, ammunition, 192, 193, 195, 197-199.
Light-boxes, 189.
Liman, battle of, 16.
Line of sight, 22, 23.
Liner, 82, 105, 106.
Lion, 342.
Lloyd, Commander, 14, 66, 151, 183
Loading, 7, 8, 49, 78, 81, 82, 131-134, 163, 164, 193-200; central, 196-199; hydraulic, 133, 134; tray, 131, 132, 133, 192.
Locking bolt, 134.
Locks, percussion, 19-21, 180-182.
Longridge, Dr., 106, 107, 347-349.
Lubricants, 139, 225.
Luting, 191.

Machiavelli, 263, 264.
Machine guns, 159-163.
Magazines, 188-193; cases, 190-192.
Magenta, 320, 321.
Magnificent, 314-316.
Maitland connection, 92, 93; groove, 55.
Majestic, 200, 314-316.
Manby, Captain, 240.
Mandril, 59.
Manganese, 97, 99; bronze, 136.
Marceau, 320, 321.
Marie de la Tour, 1.
Marlborough, 18.
Mars, 314-316.
Martin's shell, 237, 238.
Mary Rose, 4.
Mascolo or Moscolo, 8.
Massena, 322, 323
Match, 14, 19.
Maxim gun, 160-163.
Melinite, 256.
Merrimac, 296-299.
Mersey, 18.
Mill cake, 218.
Minions, 6.
Mitrailleuse, 159.
Mixing ingredients of gunpowder, 218.
Monarch, 303, 304.
Monitor, 298-300
Monson, Sir William, 6, 9.

Moore, Sir Jonas, 8, 12, 22.
Moors of Tunis, 1.
Moorsom, Captain, 30, 275, 276.
Mortar fuzes, 267, 268.
Mountings, between-deck, 142-143, 185-187; centre-pivot, 144-148; hydraulic, 76, 77, 149-158; large quick-firing guns, 172, 177, 182-187; recoil, small quick-firing guns, 160, 165-170; upper-deck, 144, 145, 185.
Murtherers, 7.

Napoleon III., 294
Nelson, Lord, 23.
Newcastle, the Duke of, 35.
Nickel steel, 350, 351.
Night sights, 207, 208.
Nile, 119, 310, 311.
Noble, Sir A., 79.
Nordenfelt guns, 159, 160, 164.
Norton, Captain, 7.

Obturation, 85, 88.
Obukoff, 352.
Oil, 76, 139.
Ores, iron, 96.
Orient, 17.
Outside bolt, 134.

Pad, De Bange, 87, 88, 127, 130.
Paixhans, Colonel, 16, 17.
Palliser, Sir W., guns, 50, 51; projectiles, 248, 249.
Peirse, Lieutenant, 131.
Percussion firing gear, 179-182; fuzes, 274-291; tubes, 19, 20, 292.
Petrieroes, 7, 8.
Pettman's fuze, 277, 279.
Pig-iron, 96, 97.
Pipes, hydraulic, 152.
Polygroove rifling, 36, 83.
Ports in turrets, 77.
Powder (*see* Gunpowder); cases, 190-192.
Powerful, 126.
Pozzuoli, 352.
Primers, electric, 179.
Priming horn and iron, 14.
Prince George, 314-316.
Prismatic powder, 221, 224.
Products of combustion of powder, 216.
Projectiles, 16, 32, 40, 229-262; armour-piercing, 247, 253; Armstrong, 40, 244-246; elongated, 48, 49, 241; Holtzer, 249; Painting, 242; Palliser, 248, 249; spherical, 34, 35, 47, 48, 229-240; stowage, 193.
Pumps, washing out, 133.
Purchases, hydraulic, 192.

Quick-firing guns, 159-187; loading, 163.
Quoin, 13, 26.

Rabinet, 6.
Racers, 66, 70, 72, 73.
Raleigh, Sir Walter, 9.
Ramilies, 119, 314-316.
Rammer, hydraulic, 193, 194
Ramming home, 195
Rangoon oil, 76, 139.
Rapidity of fire, 123, 170, 171, 172; of loading, 163.
Recoil, in line of fire, 177, 182; press, 137, 140, 141, 147, 154, 183; hydraulic mountings, 137-139, 153-157; quick-firing mountings, 160; 3-pr. and 6-pr

168, 170; valves, 137-138, 153-156.
Regent, 2, 3.
Renown, 117.
Repulse, 119, 314-316.
Requin, 318.
Resolution, 119, 314-316.
Revenge, 119, 314-316.
Rifled guns, 32, 33-95.
Rifling, guns, 62, 103; increasing twist, 53, 54, 55; polygroove, 36; shunt, 51; systems of, 49; twist of, 53, 54, 55; uniform, 53, 54, 55.
Roberts, John, 13.
Robins, Benjamin, 32.
Rodney, 119, 308, 309.
Roller paths, 147, 148; rings, 148, 149.
Rollers, 68, 72, 73, 146, 147.
Rough boring, 58, 99.
Royal Oak, 119, 311.
Royal Sovereign, 119, 192, 313.
Ruelle, 352.
Ruggiero di Lauria, 121.

Saker, or sacar, 5, 6
Saltpetre, 216, 217.
Sans Pareil, 117, 120.
Schwartz, Friar, 215.
Scott, Rear-Admiral R, 73, 74, 75.
Screwing, connection by, 102; the breech, 84, 92.
Screws, breech, 39, 84, 85, 88, 126-133, 177, 178, 180.
Separate loading, 192.
Serrations, 103.
Severn, 114.
Shafts or trunks, ammunition, 192, 193, 197, 198, 199.
Shells, 16, 17, 34, 233-238, 244-246, 249-251, 254-257, 294; bursters, 234-237, 245, 246, 248, 250, 251, 254, 255, 256; common, 234, 246, 254; double, 250; rooms, 192, 193; segment, 245; shrapnel, 235-237, 250, 254.
Shrapnel, General, 235.
Shrinkage, 90.
Shrinking on hoops and jackets, 37, 60, 61, 101, 102.
Siemens-Martin process, 96.
Sight clamp, Stuart, 203, 204; strips, 203.
Sighting gear, 203, 204; port and stations, 157, 206.
Sights, 22, 23, 24, 41-43, 200-209; Armstrong, 41-43; dispart, 22, 23; H-speed, 201, 202; night, 207, 208; quick-firing guns, 206
Silk cloth, 22.
Silloth, 172
Skoda, 352.
Slides, gun carriage, 66-74, 141-144, 153-157, 204; inclination of, 144, 146.
Sinope, 324, 325.
Smokeless explosives, 223-228.
Sovereign, 2, 3.
Sovereign of the Seas, 9, 10.
Spherical projectiles, 47, 48, 229-240.
Sponsons, 145, 318.
Springs, running out, 182, 183.
St. Louis, 323, 325.
St. Vincent, 342.
Stanhope lever, 85, 128.
Steel, 89-91, 96-98; barrels, 57, 63, 64, 89; cast, 96, 97, 98; forgings, 96, 97, 98; soft, 91.

Stowage of gunpowder, 188-192, 222.
Strains, internal, 58, 94, 99, 100, 102.
Strength, circumferential, 108.
Stuart sight clamp, 203, 204.
Sulphur, 217; purification of, 217.
Swedish pig iron, 96

Tappet gear, 39, 40.
Tartaglia, Nicolas, 5.
Tchesme, 324, 325.
Tempering steel, 58, 90.
Terni, 352.
Terrible, 126, 198, 314, 318.
Test pieces, 57, 99.
Thunderer, 64, 76, 117, 149, 150, 304.
Time and concussion fuzes, 287, 288.
Time fuzes, 266-274, 287-291.
Toulla, 352.
Trafalgar, 113, 119, 310
Training gear, 69, 72, 73, 142, 144, 147; quick-firing mountings, 184.
Trials, 4·7-inch quick-firing guns, 170, 171; 6-inch quick-firing guns, 172-174.
Truck carriages, 12, 24-27, 135.
Trunnion blocks, 187.
Trunnions, 24, 59, 60, 113-118, 168, 169, 183, 187.
Tubes, 19, 20, 180; electric, 205; friction, 292; vent-sealing, 293.
Turret mountings, 76, 77, 149-158.

Uniform rifling, 36, 53, 54.
Upper-deck and between-deck mounting, 143, 185, 186, 187.

Valve, bye-pass, 140, 141, 144; key, 138, 139.
Vavasseur, Mr., 137-146; recoil press, 137-141; valves, 137-139, 146.
Venn, Captain, 10, 264. 20
Vent-sealing tubes, 293.
Victoria, 120, 121.
Victorious, 314-316.
Victory, 15.

Wahrendorff, Baron, 33.
Warrior, 295.
Washing-out pump, 133.
Whitworth, Mr., 43-45, 98, 352.
Windage, 40, 45, 49, 56.
Wire guns, 95, 106-111, 125, 126, 174-176; winding tension, 110.
Witkovitz, 352
Woodbridge, Dr., 106.
Woolwich, 4, 46, 49, 53, 55, 89, 95, 102, 129, 137, 352.

CHISWICK PRESS:—CHARLES WHITTINGHAM AND CO
TOOKS COURT, CHANCERY LANE, LONDON.

Made in United States
Cleveland, OH
24 October 2026

46816149R00224